Berichte aus dem Institut für Mehrphasenström

Band 3

Experimental analysis of fast reactions in gas-liquid flows

Vom Promotionsausschuss der
Technischen Universität Hamburg

zur Erlangung des akademischen Grades

Doktor-Ingenieur (Dr.-Ing.)

genehmigte Dissertation

von
Jens Timmermann

aus
Höxter

2018

Bibliografische Information der Deutschen Nationalbibliothek

Die Deutsche Nationalbibliothek verzeichnet diese Publikation in der Deutschen Nationalbibliographie; detaillierte bibliographische Daten sind im Internet über http://dnb.d-nb.de abrufbar.

1. Aufl. - Göttingen: Cuvillier, 2018

   Zugl.: (TU) Hamburg-Harburg, Univ., Diss., 2018

1. Gutachter: Prof. Dr.-Ing. Michael Schlüter

2. Gutachter: Prof. Dr. rer. nat. Andreas Liese

3. Gutachter: Prof. Dr.-Ing. Hans-Joachim Warnecke

Vorsitzender des Prüfungsausschusses: Prof Dr. rer. nat. Raimund Horn

Tag der mündlichen Prüfung: 13. September 2018

© CUVILLIER VERLAG, Göttingen 2018

   Nonnenstieg 8, 37075 Göttingen

   Telefon: 0551-54724-0

   Telefax: 0551-54724-21

   www.cuvillier.de

   ISBN 978-3-7369-9875-9

   eISBN 978-3-7369-8875-0

Vorwort

Die vorliegende Arbeit entstand während und kurz nach meiner Tätigkeit als wissenschaftlicher Mitarbeiter am Institut für Mehrphasenströmungen an der Technischen Universität Hamburg, wo ich von Oktober 2014 bis Dezember 2017 ein Teilprojekt des von der Deutschen Forschungsgemeinschaft (DFG) geförderten Schwerpunktprogramms "Reaktive Blasenströmungen"[1] bearbeitet habe. Wesentliche Inhalte dieses Projektes haben die Grundlage für diese Arbeit gebildet. Bei der DFG bedanke ich mich für die finanzielle Förderung.

Mein besonderer Dank gilt meinem Doktorvater, Herrn Professor Dr.-Ing. Michael Schlüter, für das Ermöglichen dieser Arbeit, für seine Unterstützung und sein großes Interesse an dieser Fragestellung. Ohne die zahlreichen Anregungen, die wertvollen Ratschläge und die interessanten und umfangreichen Diskussionen wäre diese Arbeit in dieser Form nicht enstanden.

Herrn Professor Dr.-Ing. Hans-Joachim Warnecke danke ich für die Übernahme des Koreferats, die vielen Anregungen und passenden Ratschläge und im besonderen für die Vermittlung an das Institut für Mehrphasenströmungen.

Ebenso möchte ich Herr Professor Dr. rer. nat. Andreas Liese für die Übernahme des Koreferats und die zahlreichen Diskussionen im Rahmen des Forschungsschwerpunktes danken. Herrn Professor Dr rer. nat. Raimund Horn danke ich für die Übernahme des Prüfungsvorsitzes.

Allen ehemaligen Kolleginnen und Kollegen des Instituts für Mehrphasenströmungen ein herzliches Dankeschön. Die enge Zusammenarbeit, die nahezu zahllosen interdisziplinären Diskussionen und die gegenseitige experimentelle sowie theoretische Unterstützung hat einen wesentlichen Beitrag zu dieser Arbeit geleistet. Auf unseren Zusammenhalt, auch in turbulenten Zeiten, konnte ich mich stets verlassen.

Ein großes Dank gilt den Studierenden, deren Studien-, Bachelor- und Master-Arbeiten ich betreuen durfte, sowie den hilfswissenschaftlichen Mitarbeitern die mich während des Anlagenaufbaus und bei den zahlreichen Messkampagnen unterstützt haben.

---

[1] Schwerpunktprogramm SPP 1740 "Einfluss lokaler Transportprozesse auf chemische Reaktionen in Blasenströmungen " (FKZ: SCHL 617/12-1) der Deutschen Forschungsgemeinschaft

Den 17 Projektpartnern des SPP 1740 "Reaktive Blasenströmungen" danke ich für die Zusammenarbeit, den regen Austausch und die gemeinsamen Publikationen. Die interdisziplinäre Zusammenarbeit hat mich für viele unterschiedliche Fragestellungen und Ansätze sensibilisiert und war mir stets ein zusätzlicher Ansporn.

Meiner Familie und meinen Freunden danke ich für den Rückhalt auf den ich mich stets verlassen konnte und vor allem für das Verständnis während der "heißen" Endphase, bei der mir häufig schlicht die Zeit fehlte.

Dir Nadine, für deine schier endlose Geduld und das liebevolle Verständnis, wenn du mich oftmals vermissen musstest, für den Rückhalt und die Kraft diese Arbeit zu vollenden, ein besonderes Dankeschön.

Hamburg, im September 2018

Für Eleonore

Contents

List of Figures

List of Tables

Nomenclature

Roman Symbols

A	m^2	surface area
a	$\mathrm{m}^2 \cdot \mathrm{m}^{-3}$	specific interfacial area
C	-	constant
c_g	$\mathrm{mol} \cdot \mathrm{m}^{-3}$	concentration within the gaseous bulk phase
c_l	$\mathrm{mol} \cdot \mathrm{m}^{-3}$	concentration within the liquid bulk phase
c_v	$\mathrm{mol} \cdot \mathrm{m}^{-3}$	concentration of the liquid
$c_{\mathrm{Na_2SO_3}}$	$\mathrm{mol} \cdot \mathrm{m}^{-3}$	sodium sulfite concentration
$c_{\mathrm{O_2}}$	$\mathrm{mol} \cdot \mathrm{m}^{-3}$	oxygen concentration
c_A	$\mathrm{mol} \cdot \mathrm{m}^{-3}$	concentration of reactant A
c_{g*}	$\mathrm{mol} \cdot \mathrm{m}^{-3}$	concentration at the interface at gaseous side
c_{l*}	$\mathrm{mol} \cdot \mathrm{m}^{-3}$	concentration at the interface on liquid side
c_B	$\mathrm{mol} \cdot \mathrm{m}^{-3}$	concentration of reactant B
c_{bulk}	$\mathrm{mol} \cdot \mathrm{m}^{-3}$	concentration within the bulk phase
c_P	$\mathrm{mol} \cdot \mathrm{m}^{-3}$	concentration of product P
D	$\mathrm{m}^2 \cdot \mathrm{s}^{-1}$	diffusion coefficient
d_b	m	bubble diameter
d_p	m	pipe diameter
$D_{\mathrm{O_2}}$	$\mathrm{m}^2 \cdot \mathrm{s}^{-1}$	diffusion coefficient of oxygen in water
$D_{\mathrm{SO_3^{2-}}}$	$\mathrm{m}^2 \cdot \mathrm{s}^{-1}$	diffusion coefficient of sulfite in water
$D_{\mathrm{SO_4^{2-}}}$	$\mathrm{m}^2 \cdot \mathrm{s}^{-1}$	diffusion coefficient of sulfate in water

D_A	$\text{m}^2 \cdot \text{s}^{-1}$	diffusion coefficient of reactant A
D_B	$\text{m}^2 \cdot \text{s}^{-1}$	diffusion coefficient of reactant B
$d_{b,eq}$	mm	bubble equivalent diameter
D_C	m	column diameter
$d_{eq,A}$	mm	surface equivalent bubble diameter
$d_{eq,V}$	mm	volume equivalent bubble diameter
E	-	theoretical Enhancement factor
E^*	-	experimental Enhancement factor
F_b	$\text{kg} \cdot \text{m} \cdot \text{s}^{-2}$	buoyancy force
F_d	$\text{kg} \cdot \text{m} \cdot \text{s}^{-2}$	drag force
f_N	s^{-1}	natural frequency of oscillation
g	$\text{m} \cdot \text{s}^{-2}$	gravitational constant
H	$\text{m}^3 \cdot \text{Pa} \cdot \text{mol}^{-1}$	Henry constant
h_b	m	distance between bubble interfaces
H_C	m	vortex diameter
i	-	number
c_d	-	drag coefficient
I	-	intensity
I_0	-	reference intensity
J	$\text{mol} \cdot \text{m}^{-2} \cdot \text{s}^{-1}$	amount of substance density
k_L	$\text{m} \cdot \text{s}^{-1}$	liquid side mass transfer coefficient
k_G	$\text{m} \cdot \text{s}^{-1}$	gaseous side mass transfer coefficient
k_L^R	$\text{m} \cdot \text{s}^{-1}$	liquid side reactive mass transfer coefficient
K_{SV}	-	Stern-Volmer constant
l	m	length
$\dot{M}$	$\text{kg} \cdot \text{m}^{-3} \cdot \text{s}^{-1}$	mass flow rate
M	$\text{kg} \cdot \text{mol}^{-1}$	molar mass
n	mol	amount of substance

n_i	mol	amount of substance for an arbitrary species
n_p	mol	amount of substance for the product species
n_{ref}	mol	amount of substance for a reference species
p	$kg \cdot m^{-1} \cdot s^{-2}$	pressure
R	$kg \cdot m^2 \cdot s^{-2} \cdot K^{-1} \cdot mol^{-1}$	universal gas constant, 8.314
r	m	radius
r_b	mm	bubble radius
S_p	-	selectivity
T	K	temperature
t	s	time
t_D	s	diffusion time
t_R	s	relaxation time
V_B	m^3	bubble volume
V_R	m^3	reactor volume
$\dot{V}$	$m^3 \cdot s^{-1}$	volume flow rate
w	$m \cdot s^{-1}$	velocity
w_b	$m \cdot s^{-1}$	bubble rise velocity
$w_{b,abs}$	$m \cdot s^{-1}$	absolute bubble rise velocity
$w_{b,rel}$	$m \cdot s^{-1}$	relative bubble rise velocity
w_g^0	$m \cdot s^{-1}$	superficial gas velocity
w_l	$m \cdot s^{-1}$	liquid velocity
x	m	spatial coordinate in x direction
Y_p	-	yield
y	m	spatial coordinate in y direction
X_i	-	conversion
z	m	spatial coordinate in z direction
z_{GRID}	m	computational grid size

Nomenclature

Dimensionless numbers

$$Eo = \frac{g\Delta\rho d_b^2}{\sigma}$$ Eötvös number

$$Fr = \frac{w_b^2}{gd_p}$$ Froude number

$$Ha = \sqrt{\frac{t_D}{t_R}}$$ Hatta number

$$Mo = \frac{g\eta^4\Delta\rho}{\rho^2\sigma^3}$$ Morton number

$$Re = \frac{w_b d_b \rho}{\eta_l}$$ Reynolds number

$$Sc = \frac{v}{\rho D}$$ Schmidt number

$$Sh = \frac{d k_L}{D}$$ Sherwood number

$$We = \frac{\rho_l w_b^2 r_b}{\sigma}$$ Weber number

Greek Symbols

δ	m	boundary layer thickness
δ_g	m	gas side boundary layer thickness
δ_l	m	liquid side boundary layer thickness
$\delta_{l,reac.}$	m	reactive liquid side boundary layer thickness
η	$\mathrm{kg m^{-1} \cdot s^{-1}}$	viscosity
η_l	$\mathrm{kg m^{-1} \cdot s^{-1}}$	liquid viscosity
v	-	stoichiometry factor
v_p	-	stoichiometry factor for the product
v_{ref}	-	stoichiometry factor for a reference
ρ	$\mathrm{kg \cdot m^{-3}}$	density
ρ_g	$\mathrm{kg \cdot m^{-3}}$	gas density
ρ_l	$\mathrm{kg \cdot m^{-3}}$	liquid density
σ	$\mathrm{N \cdot m}$	surface tension
Θ		stagnant cap angle
ε_g	-	gas holdup

Superscripts

a   exponent a

b   exponent b

Acronyms / Abbreviations

CLSM   confocal laser scanning microscopy

DFG   Deutsche Forschungsgemeinschaft

DI   deionized

DMSO   dimethyl sulfoxide

LED   light emitting diode

NaN   not a number

NMR   nuclear magnetic resonance

p-LIF   planar laser induced fluorescence

ROI   region of interest

SFM   SuperFocus-Mixer

SPP   Schwerpunkt Programm

surfactant   surface active agent

TRS-LIF   time resolved scanning laser induced fluorescence

Abstract

Reactive bubbly flows are widely used in industrial processes to perform fast gas-liquid reactions, although the investigations are already intensified in the 1970s, a reliable prediction of yield and selectivity for fast parallel/consecutive reactions is still not possible. In this work different benchmark setups are developed to investigate the influence of mixing on mass transfer and chemical reaction. These setups will allow the investigation of a parallel/consecutive reaction in the same manner in the future.

For a reliable prediction of the yield and selectivity, the accurate determination of the intrinsic kinetics without mass transfer limitation is required. An experimental setup based on the SuperFocus-Mixer (SFM) is developed and successfully used to determine the kinetics of the oxidation of sodium sulfite in combination with numerical simulations.

The knowledge of the intrinsic kinetics allows the investigation of the reactive mass transfer in bubbly flows in more detail. As a first step, todays techniques are applied to a known problem, the rectilinear bubble rise (bubble diameter d_b below 1 mm), to gain data as a reference for numerical simulations. Planar laser induced fluorescence (p-LIF) is applied to determine local concentration fields, which enables the evaluation of the mass transfer coefficient through mass balancing. The results of rise velocity and mass transfer coefficient testify a surfactant behavior of the fluorophore, which is so far not considered within the literature.

In a second step, with a higher degree of complexity, the physical and reactive mass transfer in case of bubble-bubble interactions is investigated. By the use of background illumination recording, the mass transfer coefficient in dependency of the bouncing frequency is quantified and key effects for mass transfer enhancement are identified by analysis of local concentration fields. Based on the findings, a semi-empirical model for the description of the determined dependency is developed.

For the development of parallel/consecutive reaction systems for gas/liquid flows, that are applicable under academic conditions, a setup with small volume is required, which allows detailed investigations at different mixing conditions already in a small volume.

The system of choice is the Taylor bubble experiment that has been developed within the DFG SPP 1740 with a few adaptations for oxygen exclusion. With different substance and reaction systems it is proven that the results are transferable and reliable.

Because the wake structure of free rising bubbles above d_b <1 mm typically not show a rotational symmetry, within this work the time resolved scanning laser induced fluorescence (TRS-LIF) technique is used to investigate the three-dimensional concentration field. Additionally, separation points are easily identified by this technique, so that significant data for numerical validations are supplied.

This thesis contributes within the SPP 1740 "Reactive Bubbly Flows" to a better understanding of gas-liquid reactions in bubbly flows. Within this framework, several cooperations are used to develop an approach for the detailed investigation of parallel/consecutive reactions. Additionally, new insights in mass transfer processes at rectilinear rising bubbles, bubble-bubble interactions and wake structures are obtained. Furthermore, a model for the description of bubble bouncing is developed, which can be included in numerical simulations of bubble columns.

Zusammenfassung

Obwohl reaktive Blasenströmungen bereits vielfach in der chemischen Industrie zur Durchführung von gas-flüssig Reaktionen verwendet werden und die Untersuchung derselben bereits mit dem Beginn der 1970er Jahre intensiviert wurden, ist die Vorhersage von Ausbeute und Selektivität schneller, paralleler oder konsekutiver Reaktionen bisher noch nicht möglich.

Um eine belastbare Vorhersage von Ausbeute und Selektivität chemischer Reaktionen zu erreichen, ist eine genaue Bestimmung der intrinsischen Kinetik ohne Stofftransport Limitierung erforderlich. In dieser Arbeit wurde daher ein Versuchsaufbau basierend auf dem SuperFocus Mischer (SFM) entwickelt und erfolgreich eingesetzt, um die Kinetik der Sulfitoxidation zu bestimmen.

Die Kenntnis der intrinsischen Kinetik erlaubt den reaktiven Stofftransport in Blasenströmungen detailliert zu untersuchen. Dazu wird in einem ersten Schritt ein bereits bekanntes Problem, der geradlinige Blasenaufstieg (Blasendurchmesser d_b unter 1 mm), detailliert betrachtet und mit numerischen Ergebnissen validiert. Zu diesem Zweck werden unter Verwendung der planaren Laser Induzieren Fluoreszenz (p-LIF) lokale Konzentrationsfelder aufgezeichnet und der Stofftransportkoeffizient ermittelt. Die Ergebnisse der Blasenaufstiegsgeschwindigkeit und Stofftransportkoeffizienten belegen, dass der verwendete Fluoreszenzfarbstoff als oberflächenaktive Substanz fungiert, was in der aktuellen Literatur bisher jedoch nicht berücksichtigt wird. Basierend auf diesem Ergebnis werden von der Arbeitsgruppe Bothe an der TU Darmstadt numerische Simulationen durchgeführt, welche die experimentellen Ergebnisse ebenfalls bestätigen.

In einem zweiten Schritt, mit einer erhöhten Komplexität, wird die Auswirkung von Blase-Blase-Interaktionen auf den physikalischen und reaktiven Stofftransport untersucht. Mithilfe von Gegenlichtaufnahmen wird zunächst der Stofftransportkoeffizient in Abhängigkeit der Stoßfrequenz ermittelt und anschließend basierend auf diesen Ergebnissen ein semiempirisches Modell zur Beschreibung dieser Abhängigkeit entwickelt. Um Schlüsselschritte zu identifizieren, erfolgt eine Bestimmung lokaler Konzentrationsfelder mittels p-LIF. Die beobachteten Phänomene werden anschließend in einer modellhaften Beschreibung zusammengefasst.

Zur Entwicklung eines parallelen bzw. konsekutiven Reaktionssystems in gas/flüssig Strömungen bei akademischen Bedingungen, muss ein Versuchsaufbau gefunden werden, der eine detaillierte Untersuchung bei unterschiedlichen Mischungszuständen und gleichzeitig kleinem Probenvolumen erlaubt. Das Taylor Blase Experiment, welches innerhalb des DFG SPP 1506 entwickelt wurde, erscheint mit kleineren Änderungen, wie beispielsweise dem Ausschluss von Sauerstoff, als geeignet. Durch Versuche mit unterschiedlichen Reaktions- und Stoffsystemen wird gezeigt, dass die Ergebnisse dieses Versuchsaufbaus belastbar und eine Übertragung auf allgemeine Blasenströmungen möglich ist.

Ellipsoide und formdynamische Blasen, wie sie in realen Blasenströmungen auftreten, zeigen jedoch keine so gut einstellbare Nachlaufstruktur wie Taylor Blasen oder eine Symmetrie wie kleine

Blasen (d_b <1 mm). Daher muss eine Technik gefunden werden, die es erlaubt auch diese komplexen Nachlaufgebiete zu visualisieren. Im Rahmen der vorliegenden Arbeit wird dazu die Time Resolved Scanning Laser Induzierte Fluoreszenz (TRS-LIF) weiterentwickelt und eingesetzt, um dreidimensionale Konzentrationsfelder zu untersuchen. Zusätzlich erlaubt diese Technik auch das Aufzeigen von Separationspunkten und liefert daher wichtige Daten für eine numerische Validierung.

Die vorliegende Arbeit wird im Rahmen des SPP 1740 „Einfluss lokaler Transportprozesse auf chemische Reaktionen in Blasenströmungen" mit dem Ziel eines besseren Verständnisses von gas-flüssig-Reaktionen in Blasenströmungen durchgeführt. Innerhalb dieses Netzwerks werden zahlreiche Kooperationen erfolgreich initiiert um einen experimentellen Ansatz zur detaillierten Untersuchung von parallelen und konsekutiven Reaktionen zu entwickeln. Zusätzlich werden neue Einblicke in die Stofftransportvorgänge bei geradlinigem Blasenaufstieg, bei Blase-Blase-Interaktionen und freiem Blasenaufstieg formdynamischer Blasen erreicht. Darüber hinaus wird ein Modell zur Beschreibung von Blasenkollisionen entwickelt, welches sich zur Verbesserung der numerischer Simulation von Blasensäulen eignet.

Chapter 1

Introduction

Bubbly flows are widely used within chemical process engineering to perform fast gas-liquid reactions, like for example oxidations, hydrogenations, chlorinations or alkylations. Typically the bubbles are irregularly in shape and are inducing complex flow structures within the bulk phase during the buoyancy driven rise. Until today, these complex flow structures are evading from an encompassing understanding and the influence on mixing, mass transfer and chemical reactions is as well unknown. Especially the mass transfer from the bubble into the boundary layer is tough to grasp, but has to be understood for a description of the influence on yield and selectivity of chemical reactions. The selectivity within this work is defined as

$$S_p = \frac{Y_p}{X_i} \tag{1.1}$$

Where X_i is the conversion and describes the ratio of the reacted amount of substance n_i to the total amount of reactant $n_{i,0}$

$$X_i = \frac{n_{i,0} - n_i}{n_{i,0}} \tag{1.2}$$

and Y_p is the yield which is defined as the ratio of the amount of substance of a specified product n_p and a reference amount of substance $n_{ref,0}$ under consideration of the stoichiometry (v_{ref}, v_p) [Bae87]

$$Y_p = \frac{n_p - n_{p,0}}{n_{ref,0}} \frac{|v_{ref}|}{v_p} \quad . \tag{1.3}$$

The clarification of this process is vital, since the timescales of mixing are dominating yield and selectivity through boundary layer deformations, interactions between bubbles and turbulences induced by the surrounding bubble swarm. To enable an effective utilization of the gaseous phase with a minimum energy consumption, optimal contact times and high mass transfer rates for process intensification and optimization are required and therefore a reliable and exact design of multiphase reactors have to be performed. So far empirical and semi-empirical correlations from literature or ascertained by lab/pilot plant experiments are mostly based on integral data. These correlations, like equation 1.6, usually consider the bubble diameter d_b, rise velocity w_b represented by the Reynolds number Re

1

$$Re = \frac{w_b d_b \rho}{\eta_l} \tag{1.4}$$

and substance system represented by viscosity η, diffusion coefficient D represented by the Schmidt number Sc

$$Sc = \frac{v}{\rho D} . \tag{1.5}$$

The transport resistance is dedicated as an influence of the fluid phase and the local hydrodynamic due to bubble swarms, like the mixing in chaotic wake structures, are considered within these correlations in the form of constants C_1 or exponent's a, b according to

$$Sh = 2 + C_1 \cdot Re^a \cdot Sc^b. \tag{1.6}$$

In addition to these classical descriptions, gas bubbles are considered since a few years more as a local mixing element, which allows a new control of reactive processes.

Within the priority program SPP 1740 "Reactive Bubbly Flows" of the German Research Foundation (DFG), a close interaction between engineering, chemistry and mathematics on the field of gas/liquid mass transfer with superimposed reaction is realized. This thesis contributes within the priority program to a better understanding of gas-liquid reactions.

Chapter 2

State-of-the-art

The design of gas-liquid reactors is a laborious task due to its interdisciplinary conjunction of chemical (reaction kinetics) and physical (fluid mechanics, molecular diffusion, etc.) processes. Figure 2.1 shows schematically the stages of a design process for gas-liquid reactors [Alp83].

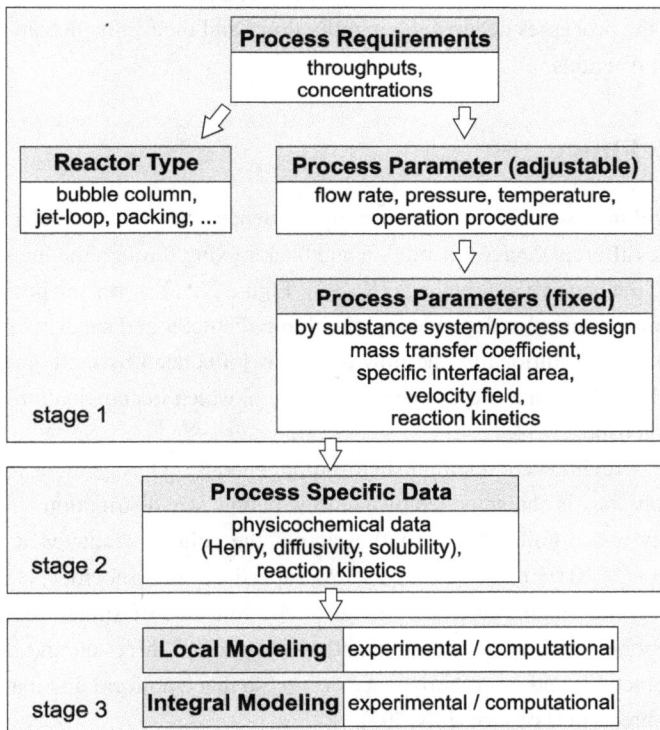

Figure 2.1 Procedure for the design of gas-liquid reactors according to Alper [Alp83].

At first the process requirements like throughput and concentrations have to be specified and lead to a variety of adjustable parameters such as temperature, pressure, flow rates or reactor type. Defining these parameters is not easy to accomplish and is today mostly based on empirical or semi-empirical correlations combined with numerical simulations. The mass transfer coefficient, the specific interfacial area, the reaction kinetics or the velocity field are typically not easily adjusted in a static geometry and are therefore fixed parameters and have to be determined within the design process.

In a second stage of process design, the fixed parameters and process specific data, like solubilities, diffusivities or reaction kinetics, are determined by experimental investigations on lab or pilot scale. These data depend strongly on the substance system and have to be determined at the operating conditions to gain a similar behavior.

Within the last stage of the design procedure, local and integral scale modeling is performed to compute the expectable performance. Several models allow already a description of flow structure or chemical reaction. Beside these complex and often inaccurate theoretical modeling, also laboratory experiments on local and/or integral scale are used. These experimental investigations often enable to bypass the experiments performed in stage two. Nevertheless, all parameters of stage one are required for a reliable design.

While a detailed procedure for designing gas-liquid reactors already exists, the prediction of yield and selectivity is still insufficient and is typically adjusted within pilot and production scale experiments. However, an improvement of energy consumption and yield is often still possible, so that a more accurate prediction is aimed by actual research and industry through gaining more detailed information about the processes occurring in bubbly flows and their influence on mass transfer and following chemical reactions.

2.1 Bubbly Flows

The flow regime within gas-liquid reactors strongly influences the yield and selectivity of chemical reactions due to the different degrees of mixing and back mixing through the interaction of bubbles with each other and the occurring turbulence [Kaš93]. Figure 2.2 a) shows the possible flow regimes within bubble column reactors in dependency of column diameter and superficial gas velocity in a watery system [Sha82]. Additionally, the flow regime is influenced by the type of gas distributer, column internals, liquid flow rate and the substance system, which in combination is the reason for a very broad transition band.

Three major flow regimes are distinguished: homogeneous, heterogeneous and slug flow. The homogeneous bubbly flow is characterized by a narrow bubble size distribution with a nearly similar bubble rise velocity and a uniform distribution across the column diameter at a low superficial gas velocity (up to $w_g^0 = 0.03$ m·s^{-1}). At very low superficial gas velocities, a small bubble size, a unidirectional rise and small velocities similar to the free rise of single bubble results. As a consequence the bubbles do not move significantly in horizontal direction and only small mixing through bubble interactions and macro turbulence occurs, so that a uniform distribution of gas holdup in radial and axial direction is observed [Kaš93].

With rising superficial gas velocity above $w_g^0 = 0.05$ m·s^{-1} the gas hold up is increasing and a turbulent flow structure with irregular distribution results.

At column diameters above approximately 0.15 m a heterogeneous bubbly flow is observed. Due to the rising gas hold up, a frequent interaction between bubble-bubble and bubble-wake occurs, so that higher shear and pressure stresses influence the bubbles and lead to break up and coalescence.

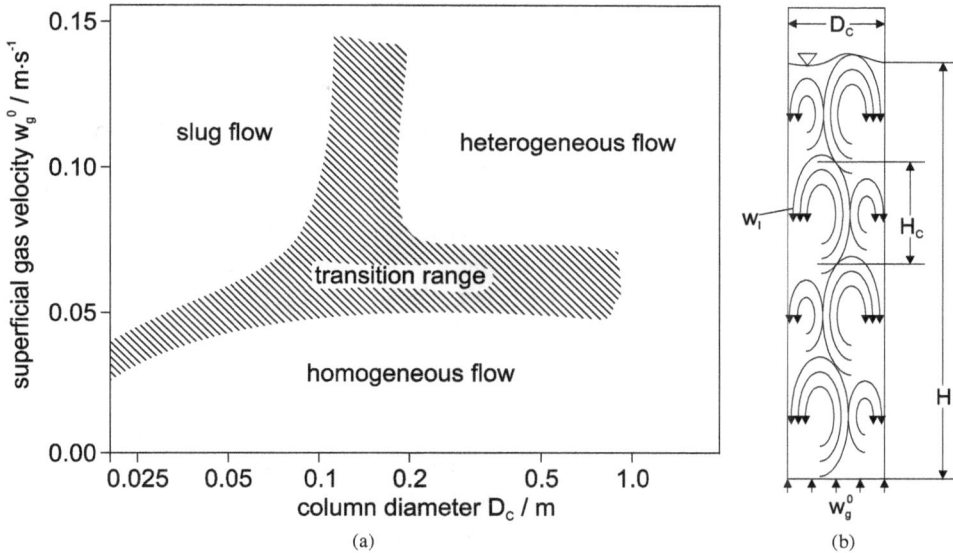

Figure 2.2 (a) Approximated dependency of flow regimes in bubble column reactors on gas velocity and column diameter in a watery system according to Shah [Sha82]. (b) Scheme of mixing zones in a bubble column reactor [Kra12].

Therefore a broad bubble size distribution with slow to fast rising bubbles is found. While the largest bubbles rise preferred in the column center with a velocity up to $1 \text{ m} \cdot \text{s}^{-1}$, the smallest bubbles are trapped in back mixing zones close to the wall [Kaš93]. Therefore, a density distribution across the column causes a downward flow close to the wall and an upwards flow in the column center, so that back mixing zones like shown in figure 2.2 b) results. The intensity of these zones depends on the superficial gas velocity, column diameter, bubble size distribution and other parameters [Kra12]. Large bubbles represent the major part of the overall gas throughput, but have a small surface to volume ratio with a small residence time, so that the contribution to mass transfer is low, but dominates the flow structure and back mixing. Nonetheless, or actually because of these observations, it is assumed that small bubbles are mainly contribute to the mass transfer performance [Kaš93].

At small column diameters below 0.15 m, which are typically used within lab scale experiments, slug flow at high gas flow rates is observed. The column diameter is here in the range of the bubble diameter, so that wall-effects are affecting the rise by stabilizing the bubble shape and lowering of the rise velocity and a characteristic slug is formed [Kra12].

2.1.1 Bubble Shape and Rise Velocity of Bubbles

The rise velocity w_b of gas bubbles within a bubble column determines several crucial parameters that affect the mass transfer performance. As already described above, the gas hold up is mainly influenced by the bubble size distribution which also determines the bubble rise velocity. Additionally, the residence time and therefore the contact time between gas and liquid phase, respectively the boundary layer thickness, are significantly influenced.

Due to the size and shape distribution additionally to the flow regime in bubble columns, a broad velocity distribution results and is typically taken into account within the design process as an averaged velocity ($\overline{w_b}$)

$$\overline{w_b} = \frac{\sum_{i=1}^{i=n} w_{b,i}}{i} \tag{2.1}$$

where n is the number of velocities. Many workers (e.g. [Nic62, Zub65, Bra71a, Uey79, Zeh85]) developed empirical correlations to describe the relative bubble velocity within a bubble swarm ($\overline{w_{b,rel}} = w_b - \overline{w_l}$) based on the rise velocity of a single bubble. In case of a stagnant bulk phase, the relative velocity is equal to the absolute velocity ($\overline{w_{b,rel}} = w_{b,abs} = \overline{w_l}$) [Sch02]. One example is the correlation according to Zehner [Zeh85]

$$\overline{w_{b,rel}} = w_b \left(1 - \varepsilon_g\right)^{m\left(1 + \frac{q\delta}{d_b}\right)^2} \tag{2.2}$$

where δ is the thickness of the velocity boundary layer and d_b the bubble diameter. By fitting this semi-empirical correlation to experimental data, the constants are set to $m = 1.75$ and $q = 3/4$. This correlation is deduced for solid spheres, but it has been shown, that the correlation is also valid for bubble swarms by Zehner [Zeh88], if the bubble velocity and bubble diameter are calculated according to Mersmann [Mer77].

Single Bubble Rise

While the influence of the flow regime on the rise velocity is already revealed, still the behavior of single bubbles is used as a parameter for the design of bubble column reactors. The rise velocity of a single bubble is determined, in the simplest case, by the size and shape through the equilibrium of buoyancy force F_b

$$F_b = d_b^3 \frac{\pi}{6} g \left(\rho_l - \rho_g\right) \tag{2.3}$$

and drag force F_d

$$F_d = c_d d_b^2 \frac{\pi}{4} \frac{\rho_l w_b^2}{2} \tag{2.4}$$

in a stagnant liquid. Identifying equation 2.3 with 2.4 and conversion to w_b lead to

$$w_b = \sqrt{\frac{4}{3} \frac{d_b g \left(\rho_l - \rho_g\right)}{c_d \rho_l}} \tag{2.5}$$

where g is the gravitational constant, ρ_l the liquid density, ρ_g the gas density and c_d the drag coefficient. However, the rise velocity in bubbly flows is typically not described that easy, due to superimposed liquid flow fields, that influence the rise velocity. The drag coefficient of a rising bubble is in difference to the drag of a solid sphere, which depends mainly on size and shape, additionally influenced by the deformation of the shape. Since a complex combination of compressibility and density gradients additionally to the interdependency between shape, size and flow around the bubble exist, a continuous description of the drag coefficient is not found so far. Figure 2.3 shows a scheme of the dependency of the drag coefficient from the Reynolds number with an additional indication for four different bubble shapes (η_l is the dynamic liquid viscosity and ρ is the density). At very low Reynolds numbers and therefore small diameters and velocities, a rising bubble behaves like a solid sphere (figure 2.3 range A). With increasing bubble diameter, the interface becomes mobile, so that internal circulations occur, which lead to a steeper decrease of the drag coefficient (figure 2.3 range B) in comparison to solid particles. Especially in this region surface-active agents (surfactants) have a high impact on the rise velocity, since the mobility of the interface is lowered. Depending on the surface coverage, the rise velocity could vary between a clean bubble and solid particles (see figure 2.4). Additionally, also the mass transfer across the interface is affected (see section 2.2.1). Within the third region (figure 2.3 range C) a rise of the drag coefficient above the level of rigid particles results due to beginning or developed deformation and oscillation of the bubbles. This rise is caused mainly by a higher pressure

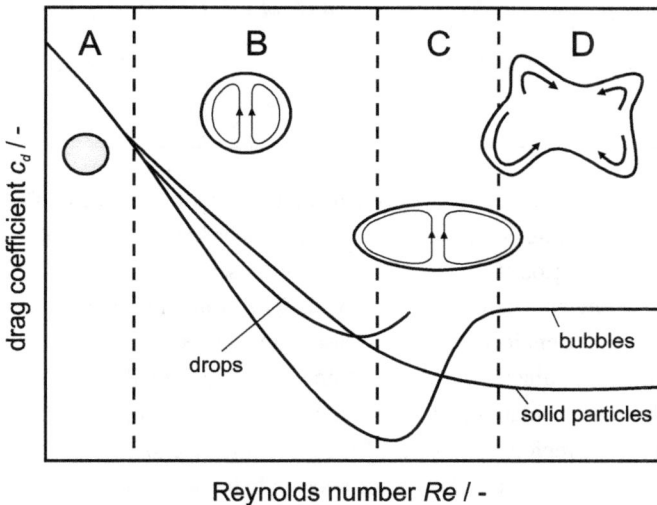

Figure 2.3 Schematic trend of the drag coefficient in dependency of the Reynolds number for solid particles and bubbles. Additional indication for four different bubble shapes [Räb13].

drop resulting from additional characteristic vortices within the bubble wake (see therefore also [Win66]). With rising Reynolds number, ring vortices or periodically forming and detaching vortices occur, that force the bubble on a zigzag, helical or irregular path. By changing the direction of movement, the rectilinear part of the rise velocity is reduced, so that the drag coefficient is increased. With a further increase of the Reynolds number, the bubble becomes irregular in shape and might disintegrate into several fragments (figure 2.3 range D) [Räb13].

Within the last decades several workers have been investigating the rise velocity of gas bubbles depending on bubble size. A good overview about the different correlations is given in [Cli78, Kul05]. So far there is no generalized correlation available to describe the rise velocity over the full bubble size range [Kul05]. Within this work only a few experimental results and correlations derived within the last decades are specified to give a brief overview.

In 1953 Peebles and Garber published a study with different correlations to describe the rise velocity for the four bubble size regions declared above. Until today this correlation is often used and cited (e.g. [Men67, Mer77, Tom98, Tom02, Kul05]). It is stated, that in the range of Re below approximately 2, spherical bubbles are moving in rectilinear paths and the drag coefficient agrees well with the results predicted by Stokes' law (table 2.1, region 1).

Table 2.1 Summarized velocity correlations for gas bubbles according to Peebles and Garber [Pee53, Kul05].

	terminal velocity	range of applicability
region 1	$w_b = \frac{2(d_b/2)^2(\rho_l - \rho_g)g}{9\eta_l}$	$Re \leqslant 2$
region 2	$w_b = 0.33^{0.76} \frac{\rho_l^{0.52} r_b^{1.28}}{\eta^{0.52}}$	$2 \leqslant Re \leqslant 4.02\,Mo^{-0.214}$
region 3	$w_b = 1.35\sqrt{\frac{\sigma g}{r_b \rho_l}}$	$4.02\,Mo^{-0.214} \leqslant Re \leqslant 3.10\,Mo^{-0.25}$
region 4	$w_b = 1.53\left(\frac{g\Delta\rho\sigma}{\rho_l^2}\right)^{0.25}$	$3.10\,Mo^{-0.25} < Re$
where	$Mo = \frac{g\eta^4\Delta\rho}{\rho^2\sigma^3}$	

Within the range of Re above 2 up to a critical value (see table 2.1) the bubbles are still moving in a rectilinear path with a drag coefficient that is slightly smaller than those of a volume equivalent solid sphere. In the third region (for range of applicability see table 2.1), the bubbles are deformed ellipsoidal and are moving on a helical or zigzag path. Therefore, the velocity in vertical direction is lowered and the drag coefficient increases. In the last region, the bubbles are already irregular in shape and are rising in a nearly rectilinear path while the drag coefficient is still increasing. Figure 2.4 shows experimental results and correlations of different work groups. Within this figure, the correlation of Peebles and Garber for region 2 is plotted. By comparing the experimental results of Duineveldt [Dui95] for clean bubbles and Peters [Pet12] for bubbles in tap water reveals, that the given equation is most likely based on experimental results with a contaminated surface.

In 1967 Mendelson [Men67] published an approach (equation 2.6) to describe the dependency of the rise velocity in region 3 and 4. This approach assumes, that large bubbles show defects at the interface with a behavior similar to waves on an ideal liquid. While the physical relation is still questionable, a good agreement with experimental data results (compare figure 2.4)

$$w_b = \sqrt{\frac{\sigma}{r_b\rho} + gr_b}. \tag{2.6}$$

Since this approach leads to an already good agreement with the experimental results, it is modified by several authors [Mar65, Com71, Cli78] to obtain a better agreement. Equation 2.7 is the modified correlation by Clift [Cli78]. According to Clift, this equation is valid in pure systems above a bubble diameter of 1.3 mm

$$w_b = \sqrt{\frac{2.14\sigma}{d_b\rho} + 0.505gd_b}. \tag{2.7}$$

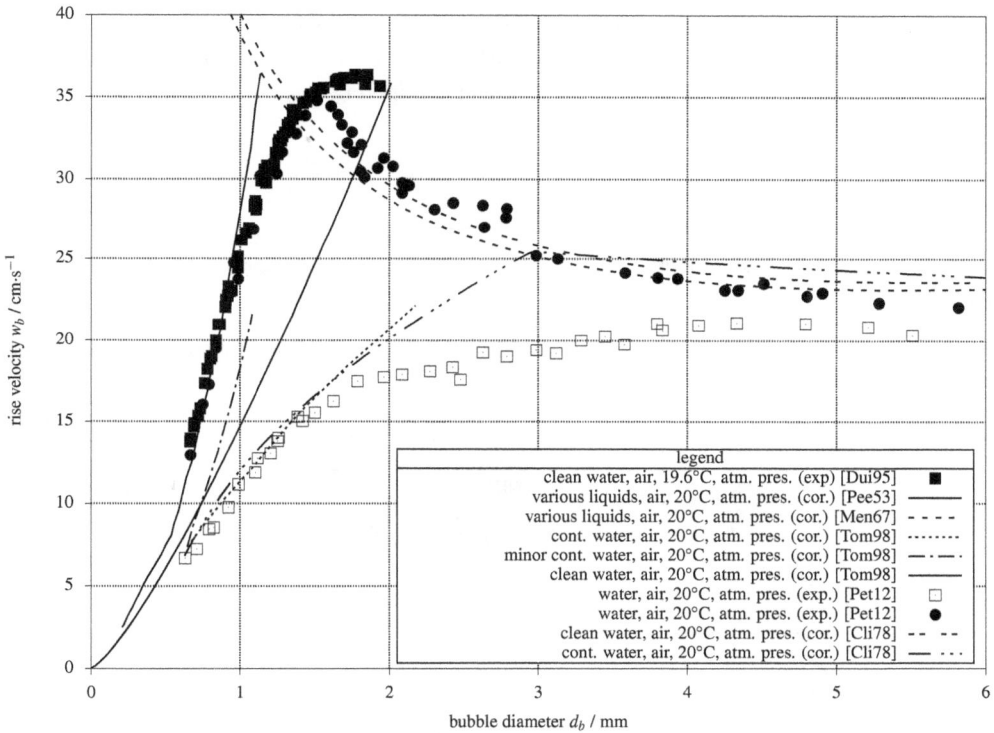

Figure 2.4 Experimental results and correlations for the rise velocity in dependency of the bubble diameter [Pee53, Men67, Cli78, Dui95, Tom98, Pet12].

To control the contamination of surfactants in aqueous solution and the preparation of surfactant free water Duineveld developed a procedure to obtain "hyper clean" water with a very small degree of contamination. The measured rise velocities are the largest measured so far in the region from 0.66 to 2 mm [Dui95]. The results are plotted in figure 2.4 and reveal a great dependency of the rise velocity within this region from the degree of contamination by comparing with the equation of Peebles and Garber.

Tomiyama investigated the rise velocity in water with several degrees of surfactant contamination. During this investigation it is found, that the generation process also has a mayor influence on the rise velocity [Tom98, Tom02]. The experimental results are modeled by the use of the drag coefficient with taking into account the effects of fluid properties, gravity and bubble diameter. The degree of contamination is separated in a pure, slightly contaminated and full contaminated system, since the degree of contamination is tough to specify. Therefore the dominating forces, drag and bouyancy are used, according to equation 2.3 to 2.5. By squaring and multiplying equation 2.5 with $(\rho_l d_b \eta_l)^2$

$$Re^2 = \frac{4}{3c_d}\sqrt{\frac{Eo^3}{Mo}} \tag{2.8}$$

results. By use of equation 2.8, the rise velocity is modeled by drag coefficient equations from other work groups with small modifications to fit the experimental results. The equations and different references are summarized in table 2.2.

Table 2.2 Drag coefficient correlations for gas bubbles and degree at different levels of contamination according to Tomiyama [Tom98].

system	drag coefficient	reference
pure	$c_d = \max\left[\min\left(\frac{16}{Re}\left(1+0.15Re^{0.687}\right), \frac{48}{Re}\right), \frac{8}{3}\frac{Eo}{Eo+4}\right]$	[Had11, Lev62, Men67]
slightly	$c_d = \max\left[\min\left(\frac{24}{Re}\left(1+0.15Re^{0.687}\right), \frac{72}{Re}\right), \frac{8}{3}\frac{Eo}{Eo+4}\right]$	[Men67, Cli71]
full	$c_d = \max\left[\frac{24}{Re}\left(1+0.15Re^{0.687}\right), \frac{8}{3}\frac{Eo}{Eo+4}\right]$	[Men67, Cli71]
where	$Eo = \frac{g\Delta\rho d_b^2}{\sigma}$	

For bubbles in regime C and D (see figure 2.3) the wave analogy of Mendelson [Men67] is applied and through non-dimensioning of equation 2.6

$$Re^2 = \frac{1}{2}\sqrt{\frac{Eo}{Mo}}\left(Eo+4\right) \tag{2.9}$$

is obtained. Afterwards equation 2.8 is identified with equation 2.9, so that

$$c_d = \frac{8}{3}\frac{Eo}{Eo+4} \tag{2.10}$$

results for the drag coefficient in this regime. With the use of equation 2.6 and the correlations for the drag coefficients, the rise velocity is calculated iteratively. The results for a water/air system are plotted in figure 2.4 [Tom98].

As already revealed by Tomiyama, the rise velocity is not only influenced by the bubble size and the degree of contamination, but also by the bubble formation process. Peters [Pet12] investigated the bubble rise in tap water with altering the bubble formation process. The experimental results are plotted within figure 2.4. By comparing the results with the experimental investigations of Duineveld, a very good agreement of the "fast" bubbles with the results in pure water is obvious. Only for bubble sizes above approximately 1.3 mm a deviation is observed, but the results agree well with the correlation of Mendelson and Clift based on the wave theory. Additional the "slow" bubbles identify remarkably good with the results of Tomiyama to a bubble size of approximately 1.3 mm.

Beside the mentioned influences, several additional parameters vary the rise velocity in bubbly flows depending on the reactor design and process control. For example also pressure gradients influence the rise velocity (e.g. pulsed bubble column) [Som13].

All explained correlations are compared in figure 2.4 for a water/air system at ambient conditions. As already described, the experimental results and correlations differ over a wide range, especially within a diameter range of 0.5 to 3 mm. The influence of surfactant contaminations and bubble generation process result in great deviations due to the beginning surface deformation and the influence of a mobile interface. In the range of very small bubbles, none or small bubble deformations and internal circulations occur, so that a modeling as rigid sphere describes the rise velocity very well. In the range of strongly deformed bubbles, respectively the spherical cap regime, which is not plotted within figure 2.4, no influence of surfactants nor bubble formation is observed [Cli78].

2.1.2 Binary Bubble Interactions

Two possible binary bubble interactions occur, coalescence and bouncing. In both cases, the bubbles approach each other until only a thin film remains. Through a further approach, the pressure within the film increases, a deformation of the interface (figure 2.5 a) and a repelling force on each bubble center results.

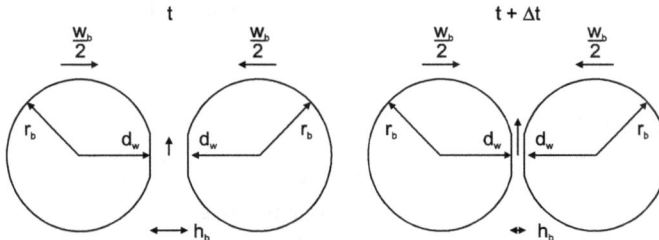

Figure 2.5 Coalescence process of two bubbles according to [Dui94].

In case of bubble coalescence the inertia lead to a further approach of the two bubbles and therefore the film thickness and velocity is decreased while the film area is increased. Therefore, the potential energy is increasing while the kinetic energy is decreased. Film drainage continues until a thickness of around 100 Å for the film is reached. At this thickness an additional force, the Van der Waals force, becomes important and results the coalescence.

In case of bubble bouncing, the bubble motion is suspended through the inertia, so that no film thinning occurs. Due to a higher approach velocity one can assume, that coalescence is preferred, but with rising velocity, also the repelling force is increased since the pressure within the film is much further increased and the motion is stopped before a film thickness of 100 Å is reached [Dui94].

Duineveld defined based on the experimental results for two bubbles rising side by side a critical Weber number, where above bouncing and not coalescence occurs

$$We_{crit} = \frac{\rho_l w_b^2 r_b}{\sigma} = 0.18 \pm 0.03 \qquad (2.11)$$

In case of water this corresponds to a velocity difference of 0.13 m·s^{-1} between the two bubbles. It is expected, that this velocity is also valid for entailing bubbles [Dui95].

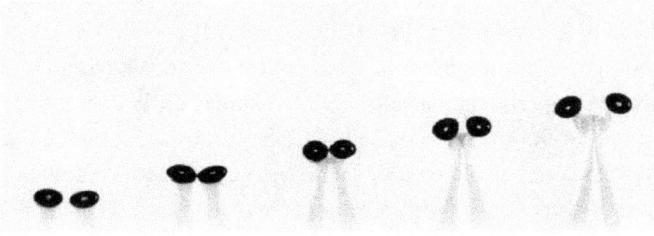

Figure 2.6 Wake of bouncing bubbles found by Sanada [San09].

Besides the influence of bubble interactions on the rise velocity and coalescence, also an influence on mass transfer is likely. Sanada [San09] investigated also the bouncing of bubbles rising side by side. For the visualization purposes, the bubbles are rising through a layer of dye solution, so that the wake is colored by the captured dye solution within the stagnation ring.

Figure 2.6 shows exemplarily the wake structure and the evolution after bouncing of both bubbles. It is observed, that the wake detaches soon after bouncing as a vortex [San09]. In case of mass transfer this results in a vortex with high concentration of transferred gas and a new wake structure is developed. Therefore, it can be assumed, that the mass transfer is influenced. This will be investigated further within this work.

2.1.3 Taylor Bubble

Bubbly flows in vertical pipes can be separated in different flow regimes as described by Bennet *et al.* in bubble, slug, churn and annular flow [Ben65]. Within this work Taylor flow is used, which is a slug flow in vertical pipes and consists of large, bullet shaped gas bubbles. Between bubble and pipe wall only a small liquid film exists, where the liquid flows downwards relative to the rising bubble. This regime begins, when the bubble diameter d_b is only slightly smaller than the pipe diameter d_p. If the bubble volume is increased, the bubble is elongated since the pipe diameter limits the cross section. Within this wall dominant regime, a nearly constant relation of bubble volume to bubble surface results for a wide range of bubble diameter, which leads to a constant rise velocity for this range [Bra71a].

Rise Velocity

The rise velocity of Taylor bubbles in water and various liquids is investigated by several authors e.g. [Dum43, Dav50, Lai56, Har60, Whi62]. White and Beardmore in 1961 [Whi62] extensively investigated the rise velocity of air Taylor bubbles in various liquids and developed based on literature and experimental results an exemplary description for the rise velocity represented by the Froude number

$$Fr = \frac{w_b^2}{g \cdot d_p} \tag{2.12}$$

where d_p is the pipe or capillary inner diameter, as a function of the Eötvös and Morton number

$$Mo = \frac{g \eta^4 \Delta \rho}{\rho^2 \sigma^3}. \tag{2.13}$$

Kurimoto et al. [Kur13] used the data of White and Beardmore to develop an empirical correlation to describe the rise velocity represented by the Froude number in dependency of the Reynolds and Eötvös number

$$Fr = \sqrt{\frac{G \cdot H}{\frac{H}{0.35^2} + F}}. \tag{2.14}$$

with

$$F = \frac{1}{Re} \left(1 - 0.05 \sqrt{Re} \right)$$

$$G = \left(a + \frac{3.8}{Eo^{1.68}} \right)^{-18.4}$$

$$H = 0.0025 \left[3 + G \right]$$

Figure 2.7 shows the experimental results of White and Beardmore as well as the results of the empirical correlation 2.14 of Kurimoto. It is observed, that there is a good agreement of the experimental data and empirical correlation, so that a good prediction of the rise velocity is possible.

An extensive overview about the progress within the last decades in description of the rise velocity of Taylor bubbles, wake structure, film thickness and shape can be found in Morgado et al. [Mor16].

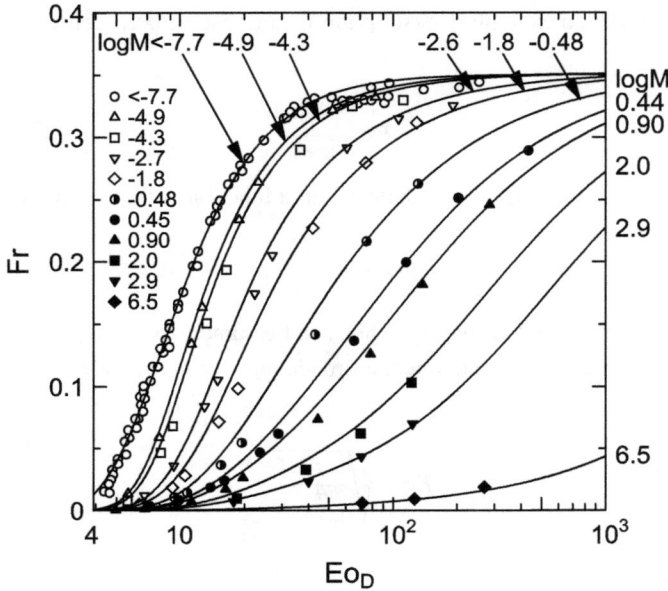

Figure 2.7 Dependency of the Froude number from the Eötvös and Morton number and experimental Data of White and Beardmore; reprinted from [Kur13].

2.2 Mass Transfer in Bubbly Flows

To describe the transfer from gas into liquid or vice versa, different models are well described within the literature (film [Lew24], penetration [Hig35] and surface renewal theory [Dan51]). For bubbly flows the film theory is still widely used to describe the mass transfer, since the more realistic, but less functional concepts of mass transfer generally not result a benefit for the exactness of gas-liquid absorber/reactor calculations [Kaš93]. The theory assumes that there is a stagnant film close to the fluid interface with the thickness δ, where the steady mass transfer takes place by molecular diffusion in one dimension normal to the interface. Additionally, it is assumed, that there is no concentration gradient within the bulk. Figure 2.8 shows a scheme of the concentration gradient within the liquid film according to the film model and the real concentration gradient, as well as the ideal mixed bulk phases outside of the film.

Lewis and Whitman described the mass transfer within the film according to Ficks's first law

$$ J = \frac{\mathrm{d}n}{\mathrm{d}t}\frac{1}{A} = -D\frac{\mathrm{d}c}{\mathrm{d}y} \tag{2.15} $$

with the amount of substance n, time t, diffusion coefficient D, the concentration c and the spatial coordinate y. Therefore a straight concentration gradient is assumed. For the amount of substance absorbed during a time step $\mathrm{d}t$, a proportional dependency from the surface of the interface A multiplied by the liquid side mass transfer coefficient k_L and the concentration gradient $(c_l^* - c_l)$ as driving force is assumed. Formulated for the liquid film

$$ \frac{\mathrm{d}n}{\mathrm{d}t\,A} = k_L \cdot (c_l^* - c_l) \tag{2.16} $$

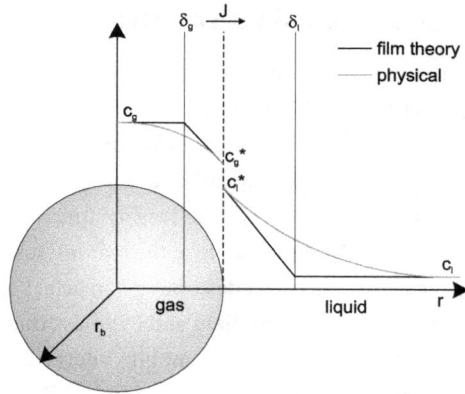

Figure 2.8 Mass transfer at a fluid particle according to the two-film theory freely adapted from [Kra12].

results. The concentration gradient within the film is described by

$$\frac{c_l - c_l^*}{c_l - c_{l,\infty}^*} = \frac{y}{\delta_l}. \tag{2.17}$$

Identifying equation 2.15 with equation 2.16, transposing to k_L and applying equation 2.17 leads to a description of the mass transfer coefficient

$$k_L = \frac{D}{\delta_l} \tag{2.18}$$

[Kra12]. Nevertheless, equation 2.18 reveals no additional information about the mass transfer process, since the film thickness δ as well as the mass transfer coefficient is in most cases unknown [Ast67]. However, on basis of the film theory an experimental determination of the mass transfer coefficient is possible and therefore the film thickness can be estimated [Kaš93]. In case of the contact of two fluid phases, like in figure 2.8 gas and liquid phase, the film theory is applied in both fluid phases and is therefore called two-film theory. The amount of substance density $J = \mathrm{d}n / (\mathrm{d}t \cdot A)$ within the gaseous phase is described by

$$J = k_G \left(c_g - c_g^*\right) \tag{2.19}$$

where k_G is the gas side mass transfer coefficient. The concentrations c_g^* and c_l^* at the interface are connected by thermodynamical relations. Typically, gases are only slightly soluble in water, so that the concentration c_l^* is generally very small. In these cases, Henry's law can be used to describe the dependency between the two concentrations

$$p = \frac{c_g^*}{c_g + c_g^* + c_v} \cdot H , \tag{2.20}$$

where p is the gas partial pressure, c_v the concentration of the liquid and H is the Henry constant [Kra12].

15

2.2.1 Influence of Surfactants on Mass Transfer

As already mentioned in section 2.1.1 beside the rise velocity, also mass transfer is affected by surfactants. This influence is described by the stagnant cap model, which proposes, that in surfactant solutions a continuous adsorption and desorption of surfactant molecules to and from the surface occurs. According to Levich [Lev62] a steady expansion of the interface at the bubble front occurs, while simultaneously the rear is compressed. On the freshly formed surface at the bubble front surfactant molecules are continuously adsorbing, while at the rear the surfactants are forced to desorb. Over time an equilibrium between adsorption and desorption is reached and a concentration gradient of surfactant on the interface arises. The velocity field and the mass transfer are directly influenced by this equilibrium between a mobile (slip) and an immobile interface (no slip, see figure 2.9). By assuming, that there is a sharp transition between mobile and immobile interface, a stagnant cap with the cap angle Θ results which leads to a mass transfer resistance [Duk15].

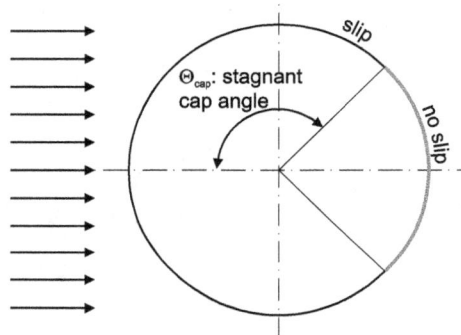

Figure 2.9 Scheme of the stagnant cap model according to [Jia17].

One model, already described by Frössling in 1938 [Fro38], for heat transfer at falling drops is still used today, to describe the mass transfer at rising bubbles with a fully immobilized surface with good agreement

$$Sh = 2 + 0.552 \, Sc^{\frac{1}{3}} \, Re^{\frac{1}{2}} . \tag{2.21}$$

Several experimental, analytical and numerical investigations have been performed to describe the influence of surfactants on droplet and bubble rise (e.g. [Gri62, Duk15]) or mass transfer performance (e.g. [Cal61]). Recently also the influence of surfactants on mass transfer [Bis91, Vas02, Alv04, Alv05, Sar06, Mad07, Küc11, Jia17] at rising bubbles gained more and more interest. For example Sardening et al. [Sar06] investigated the influence of different surfactants on rise velocity and mass transfer for oxygen bubbles in the range of 1.5 to 8 mm and developed a model to describe the mass transfer performance in this range. Therefore, the surface coverage ratio respectively the cap angle Θ (compare figure 2.9) is used to describe the change in mass transfer performance. Jia et al. [Jia17] compared their numerical results for bubbles in the same range with overall measurements of Madhavi et al. [Mad07] for the absorption of CO_2 in strong alkaline solutions. In both comparisons still a lack of agreement results and therefore a comprehensive description of the decrease of mass transfer performance is still missing. Additionally, the common considerations for the stagnant cap model are based on small bubbles with a rectilinear bubble rise, so that a lack of information for the influence of surfactants for ellipsoidal and irregular bubbles (compare figure 2.3 region 3 and 4) exists. Nevertheless, several investigations on the influence of surfactants in bubbly flows have already been performed [Álv00, Váz00, Pai05, Orh16] and revealed also a decrease in mass transfer performance. With a further improvement of predicting the mass transfer performance in case of small bubbles by numerical simulations, a better understanding of the influence in the regime of ellipsoidal and irregular bubble shape is likely.

2.2.2 Overall Mass Transfer

Typically, the overall mass transfer performance is determined by experimental investigations, where the film theory is generally applied. The mass flow rate $\dot{M}$ within the reactor can be defined as

$$\dot{M} = k_L \cdot a \cdot V_R \left(c_{m,l}^* - c_{m,l} \right) \cdot M \tag{2.22}$$

according to equation 2.16. Based on this description the mass transfer coefficient can be experimentally determined, if the specific interfacial area a is known. Nondimensionalization of equation 2.18 leads to the Sherwood number (equation 2.23), which can be used for the estimation of the mass transfer performance in bubbly flows [Bra71b]

$$Sh = \frac{d \cdot k_L}{D}. \tag{2.23}$$

Most empirical correlations for the Sherwood number are based on the Reynolds (equation 1.4) and Schmidt number Sc. According to Brauer [Bra71b] these correlations are typically in the form of

$$Sh = 2 + C_1 \cdot Re^a \cdot Sc^b. \tag{2.24}$$

where 2 is the theoretical solution for a spherical gas bubble at rest in a stagnant liquid [Kra12]. C_1 is a factor that is depending on the Schmidt number and on the ratio of the viscosities. The exponents a and b and the factor C_1 are empirically determined. It has to be stated, that these empirical correlations are only valid in a narrow range of operating conditions and typically only for the investigated gas-liquid absorber/reactor. Additionally, the applicability of these correlations is limited due to several influence parameters:

- In dependency of the bubble size, the bubble shape is periodically or non periodically deformed, so that the mass transfer in these cases is time depended.

- Furthermore, it can be expected, that turbulences close to the boundary layer are influencing the mass transfer.

- Additionally, in presence of surfactants, as already described within section 2.2.1, the surface mobility is lowered due to the surfactant coverage, which also leads to an additional mass transfer resistance.

Beside the correlations for gas-liquid absorbers/reactors also empirical and theoretical approaches for the description of mass transfer in special cases of single bubble rise exists. For example, it has been found, that shape oscillations at single bubbles lead to a mass transfer enhancement [Cli78]. One of these empirical correlations for the prediction of mass transfer in case of shape oscillations with a deviation of 6% to the experimental data was proposed by Anderson [And67]

$$Sh_e = 1.2 \left(\frac{d_{b,eq}^2 f_N}{D} \right)^{1/2}$$ (2.25)

with the equivalent diameter of the bubble $d_{b,eq}$, and the natural frequency of oscillation f_N for a single-degree-of-freedom damped system.

2.2.3 Mass Transfer at Taylor Bubbles

The mass transfer at Taylor bubbles is investigated within the last two decades extensively by several authors [VB04, Van05, Abe08, Has12, Hay14, Hos14, Kas15, Aok15]. Hosoda *et al.*, Kastens *et al.* as well as Aoki *et al.* [Hos14, Kas15, Aok15] investigated the dissolution of CO_2 Taylor bubbles in pipes with an inner diameter of 5 to 25 mm and determined the mass transfer based on the rate of bubble shrinkage. Kastens *et al.* [Kas15] developed, based on own experimental results and including results from Hosoda *et al.* [Hos14], a model for the description of the Sherwood number (equation 2.26) in dependency of the Eötvös number for CO_2 Taylor bubbles in deionized water

$$Sh_D = 290 \, Eo_D^{0.524} - \left[\frac{1.23}{(Eo_D + 1)^{0.0517}} \right]^{50.1} .$$ (2.26)

It is found, that the Sherwood number and therefore the mass transfer coefficient depend strongly on the pipe inner diameter. Additionally, it is found, that due to a constant rise velocity for a wide range of bubble diameters, in cases of small, circular pipes the mass transfer coefficient is constant (figure 2.10).

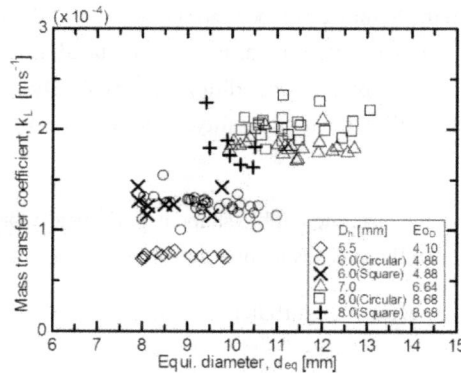

Figure 2.10 Mass transfer coefficients k_L of carbon dioxide Taylor bubbles in channels with different sizes and cross sections. (reprinted from [Kas15])

The constant rise velocity and the resulting constant mass transfer coefficient at Taylor bubbles allow quasi steady state investigations of reactive mass transfer and is used within this work for further investigations.

2.2.4 Reactive Mass Transfer

If a chemical reaction follows the mass transfer, the exemplary description becomes more complex. Due to the chemical reaction, a sink for the transferred species exists. Figure 2.11 shows exemplarily the concentration profiles according to the two-film theory (compare figure 2.8) in case of reactive mass transfer.

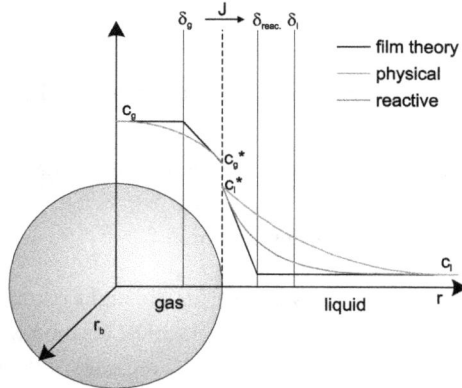

Figure 2.11 Reactive mass transfer at a fluid particle according to the two-film theory freely adapted from [Kra12].

According to the theory, the concentration of transferred gas is decreasing within the boundary layer due to the reaction faster as in the case of physical mass transfer, e.g. for a reaction of type

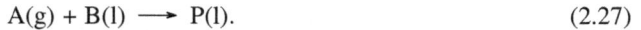

$$A(g) + B(l) \longrightarrow P(l). \tag{2.27}$$

Due to this consumption, a smaller concentration boundary layer results $\delta_{reac.}$ (see figure 2.11). The mass transfer enhancement through a chemical reaction can therefore be defined as

$$E = \frac{k_L^R}{k_L^0} = \frac{\delta_L}{\delta_L reac.} \tag{2.28}$$

based of equation 2.18, where k_L^R is the mass transfer coefficient in case of a superimposed chemical reaction. In dependency of the reaction kinetics, several concentration profiles within the boundary layer in dependency of rate constant k, the mass transfer coefficient k_G and k_L, the ratio of reaction partners c_A/c_B and the Henry constant H can occur. Based on these values, eight scenarios are derived from an instantaneous reaction (mass transfer limited) to a extremely slow reaction (no mass transfer limitation):

$\boxed{A}$ Instantaneous reaction with low c_B

$\boxed{B}$ Instantaneous reaction with high c_B

$\boxed{C}$ Fast reaction within the liquid film and low c_B

$\boxed{D}$ Fast reaction within the liquid film and high c_B

$\boxed{E/F}$ Intermediate reaction rate with reaction within the film and in the liquid bulk phase

$\boxed{G}$ Mass transfer limitation, but a slow reaction within the bulk phase

$\boxed{H}$ Very slow reaction with no mass transfer limitation

An overview about the concentration gradients for the eight scenarios is given in figure 2.12 [Lev99]. According to Levenspiel a classification of a chemical reaction within these eight scenarios is possible by using the Hatta number Ha which is generally the ratio of the relaxation time of diffusion t_D and the relaxation time $t_R = 1/k$ of a chemical reaction

$$Ha = \sqrt{\frac{t_D}{t_R}} \tag{2.29}$$

and is describing the maximum possible conversion within the film in relation to the maximum transport through the film. Since the reaction time is depending on the reaction order, different approaches result [Lev99]. According to the work of Hikita and Asai [Hik64], the Hatta number for a reaction of an $(m+n)$-th order can be described as

$$Ha = \frac{\sqrt{\frac{2}{m+1} k c_A^{m-1} c_B^n D}}{k_L} \tag{2.30}$$

where c_A and c_B are the concentrations of component A and B [Gas15]. For a reaction of first order, where $m = 1$ and $n = 0$ respectively

$$r = \frac{dc_A}{dt} = k \cdot c_A \tag{2.31}$$

for the Hatta number results

$$Ha = \frac{\sqrt{k \cdot D}}{k_L} \tag{2.32}$$

Levenspiel used the Hatta number for the classification and prediction of the expectable concentration profile within the boundary layer (see table 2.3).

With these classifications a physical description of the proposed concentration profiles within the boundary layer is possible. At very low reaction rates ($Ha < 0.02$), the concentration gradient of component A is nearly uninfluenced by the chemical reaction. With rising reaction rate ($0.02 < Ha < 2$) component A is consumed by the reaction within the concentration boundary layer, so that a steeper

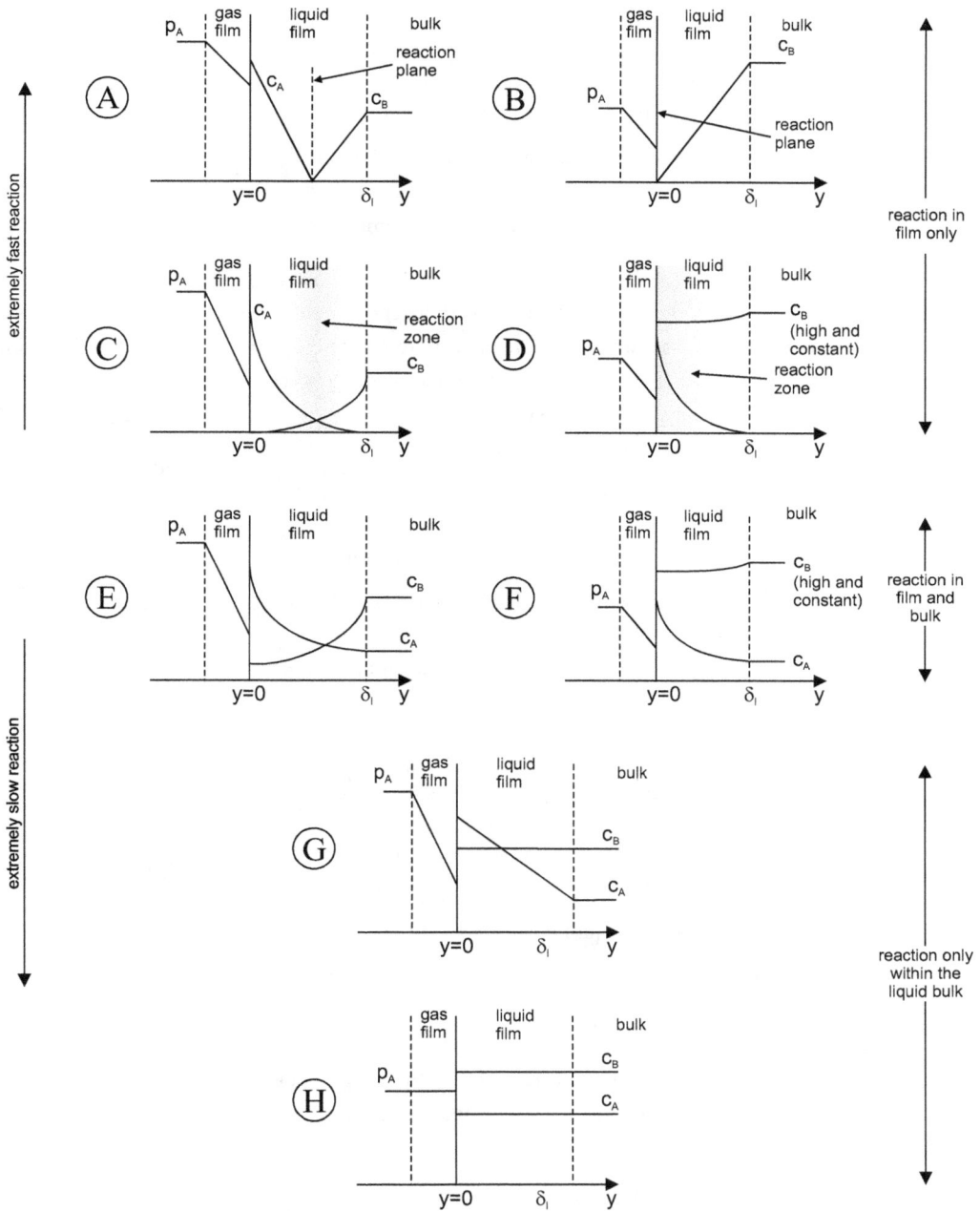

Figure 2.12 Different concentration profiles within the boundary layer in case of reactive mass transfer at a fluid particle according to the two-film theory [Lev99].

gradient results and the mass transfer between the fluid phases in accelerated. If the reaction rate becomes fast to instantaneous ($Ha > 2$) the concentration of component A is already drained within the boundary layer, so that a thinning of the concentration boundary layer results. In this regime, the

Table 2.3 Classification and prediction of the expectable concentration profiles within the boundary layer in dependency of the Hatta number according to Levenspiel (1999) [Lev99].

$Ha > 2$	Instantaneous to fast reaction within the film	$\Rightarrow$ (A), (B), (C), (D)
$0.02 < Ha < 2$	Intermediate reaction rate with reaction within the film and/or in the liquid bulk phase	$\Rightarrow$ (E), (F), (G)
$Ha < 0.02$	Very slow reaction with no mass transfer limitation	$\Rightarrow$ (H)

mass transfer rate is no longer depending on the mass transfer coefficient but on the reaction rate and diffusion coefficient [Bae87]. Nevertheless, there are different boundary conditions described within the literature for the transition between the reaction regimes and the expectable concentration profiles within the boundary layer (tables 2.4, 2.5, 2.3).

Table 2.4 Classification and prediction of the expectable concentration profiles within the boundary layer in dependency of the Hatta number according to Levenspiel (1972) [Lev72].

$Ha > 4$	Instantaneous to fast reaction within the film	$\Rightarrow$ (A), (B), (C), (D)
$0.0004 < Ha < 4$	Intermediate reaction rate with reaction within the film and/or in the liquid bulk phase	$\Rightarrow$ (E), (F), (G)
$Ha < 0.0004$	Very slow reaction with no mass transfer limitation	$\Rightarrow$ (H)

Table 2.5 Classification and prediction of the expectable concentration profiles within the boundary layer in dependency of the Hatta number according to Baerns *et al.* (1987) [Bae87].

$Ha > 3$	Instantaneous to fast reaction within the film	$\Rightarrow$ (A), (B), (C), (D)
$0.3 < Ha < 3$	Intermediate reaction rate with reaction within the film and/or in the liquid bulk phase	$\Rightarrow$ (E), (F), (G)
$Ha < 0.3$	Very slow reaction with no mass transfer resistance	$\Rightarrow$ (H)

The different authors do not explain the stated Hatta number ranges based on any literature references. It is likely, that these definitions are based on the work of Hatta [Hat32], the origin of the dimensionless Hatta number. Hatta derived for the enhancement factor E based on the condition of a fast reaction that only proceeds within the liquid film

$$E = \frac{Ha}{\tanh Ha} \tag{2.33}$$

Figure 2.13 shows a plot of equation 2.33 with the transition limits $Ha = 0.2$ from slow to intermediate reaction regime and $Ha = 3$ from intermediate to fast/instantaneous reaction regime. Hatta defined these limits based on equation 2.33 since $\tanh Ha \approx 1$ above $Ha = 3$ and also for $Ha = 0.2$ [Hat32]. Nevertheless, these arbitrarily chosen values do not represent any fixed limits within the transition of one regime to another. Based on equation 2.33, the different classification limits given within the literature can be be explained. It should be noted, that the difference can result

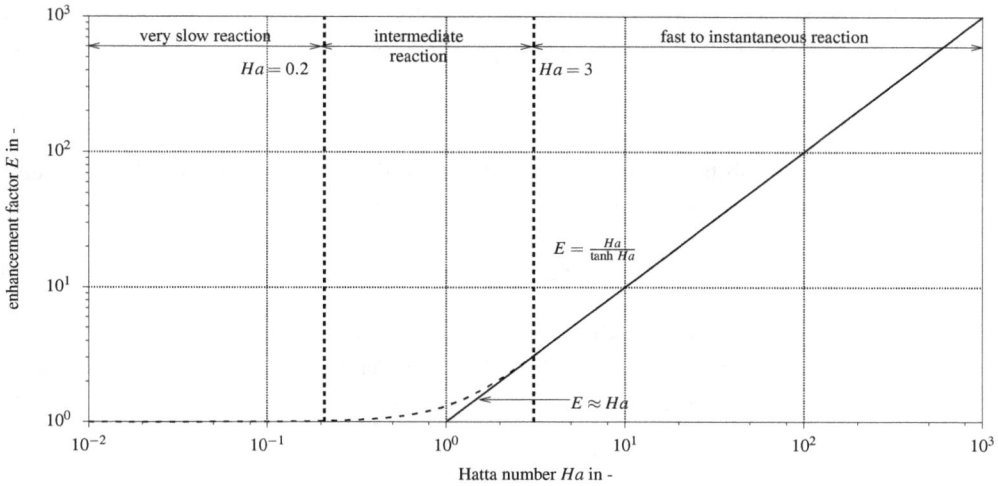

Figure 2.13 Curve shape for the function $E = Ha/\tanh Ha$ and classification of different reaction rates as stated by Hatta [Hat32].

at the upper limit nearly one order of magnitude in reaction rate and at the lower limit several orders of magnitude in reaction rate.

2.2.5 Convection Diffusion Reaction Model

To describe the mass transport of dissolved species within the liquid bulk phase, the convection diffusion reaction model (CDR-model) can be used. This model considers the convection, diffusion and chemical reaction with regard to mass conservation. Convection refers here to the forced movement of substances through the bulk phase. Diffusion is the transport of mass from one liquid element with a high concentration to a liquid element with low concentration (see 2.2). Reaction refers to a chemical reaction that is a source or sink for concentrations of substances. The model therefore describes the distribution of substances within a liquid over time under influence of these three processes.

The diffusion within the bulk phase correlates the change of concentration for dissolved species with the spacial variation in one direction

$$\frac{\partial c}{\partial t} = D \frac{\partial^2 c}{\partial x^2} \tag{2.34}$$

which is also called as the Fick's second law. This equation results directly from Fick's first law if one takes into account the net flux through the surface A and the Volume $A \cdot l$. The concentration change within this volume can be described by

$$\frac{\partial c}{\partial t} = \frac{J - J'}{l} \tag{2.35}$$

where J is the amount of substance density in the volume $A \cdot l$ and J' out of the volume. Inserting equation 2.15 and solving leads to Fick's second law (equation 2.34). The concentration change is therefore proportional to the second derivation of the concentration by location (x).

If diffusion is neglected, the transport of substance through the surface area A within a time step Δt is determined by convection with the velocity w

$$J = c \cdot w \tag{2.36}$$

This is the amount of substance density in case of a convective flux. Balancing this flux according to equation 2.35 and assuming that the velocity does not depend on the location it result [Atk06]

$$\frac{\partial c}{\partial t} = -w \frac{\partial c}{\partial x} \tag{2.37}$$

As already mentioned a chemical reaction can either be a sink or a source for substances. Therefore, the concentration change has to be written negative for reactants and positive for products as a function of the chemical reaction depending on the concentration and reaction constant

$$\frac{\partial c}{\partial t} = \pm f(c_i, k) = \pm r. \tag{2.38}$$

In case of a first order chemical reaction equation 2.31 is applied and partially written

$$r = \frac{\partial c}{\partial t} = \pm k \cdot c_A. \tag{2.39}$$

The CDR-model combines the described concentration changes

$$\frac{\partial c}{\partial t} = -w \frac{\partial c}{\partial x} + D \frac{\partial^2 c}{\partial x^2} \pm r, \tag{2.40}$$

which is a one dimensional formulation by assuming, that the diffusion coefficient and the velocity are constant. Since convection and diffusion can occur in three dimensions, a three-dimensional formulation is more likely with conservation of the balance equation structure. The three spatial coordinates are represented by x, y and z [Liu09]

$$\frac{\partial c}{\partial t} = -w \nabla c + D \nabla^2 c \pm r \tag{2.41}$$

with $\nabla c = \frac{\partial c}{\partial x} + \frac{\partial c}{\partial y} + \frac{\partial c}{\partial z}$ and $\nabla^2 c = \frac{\partial^2 c}{\partial x^2} + \frac{\partial^2 c}{\partial y^2} + \frac{\partial^2 c}{\partial z^2}$

In case of a reaction of first order, for the system of equations results

$$\frac{\partial c_A}{\partial t} = -w \nabla c_A + D \nabla^2 c_A - k \cdot c_A \tag{2.42}$$

$$\frac{\partial c_P}{\partial t} = -w \nabla c_P + D \nabla^2 c_P + k \cdot c_A \tag{2.43}$$

The CDR-model can be used to determine simultaneously the diffusion coefficients and intrinsic reaction rates of reactants with the help of a confocal laser scanning microscope (CLSM) [Bar03, War10]. Therefore, mircofluidic devices are used, which allow reproducible conditions at small substance consumption. Within these devices a stationary concentration profile results, so that

$$\frac{dc}{dt} = 0 \tag{2.44}$$

is valid. Therefore equation 2.41 simplifies to

$$w\nabla c = D\nabla^2 c \pm r, \tag{2.45}$$

which can be used to determine first the diffusion coefficient for physical mixing and afterwards the determination of the intrinsic kinetics.

2.2.6 Influence of Mixing on Mass Transfer and Reactions

It has been already shown in the literature, that mixing influences the velocity of chemical reactions [Bał99] and also that the yield and selectivity of chemical reactions are influenced by active or back mixing [Kaš93, Bał99, Bou03]. As already mentioned in the beginning, the interdependency of mixing, mass transfer and chemical reaction in case of gas-liquid reactions is not revealed yet. Nevertheless, already few regularities are known for the influence of mixing on the yield and selectivity within gas-liquid reactors. Since the character of the gas-liquid flow varies in broad limits, real bubbly flows so far are not satisfactorily described by any model [Kaš93].

As already shown in figure 2.13 chemical reactions are classified in slow, intermediate and fast to instantaneous in case of reactive mass transfer. For a consecutive reaction of type

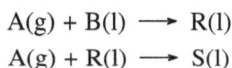

$$A(g) + B(l) \longrightarrow R(l)$$
$$A(g) + R(l) \longrightarrow S(l)$$

it is already proven that the selectivity is influenced by the interfacial mass transfer [Ter69, Ter70].

One major influence on the yield and selectivity in bubble columns is the degree of back mixing. In case of reactions with a positive order to the liquid reactant B usually back mixing lowers the reactor productivity, if the desired product is the intermediate R(l), since the selectivity is reduced, the intermediate reacts further with the gaseous reactant [Kaš93]. The effect of mixing on the selectivity in case of consecutive reactions is already investigated by Kaštánek and Fialová [Kaš82] by the use of approximating models. It is found, that the selectivity of an intermediate formation increases, if the liquid flow approaches the plug flow limit. Therefore, a bubble column with only small back mixing is required for such a reaction [Kaš93]. Nevertheless, only the homogeneous bubbly flow regime shows small back mixing, but is irrelevant for industrial applications due to the small throughput.

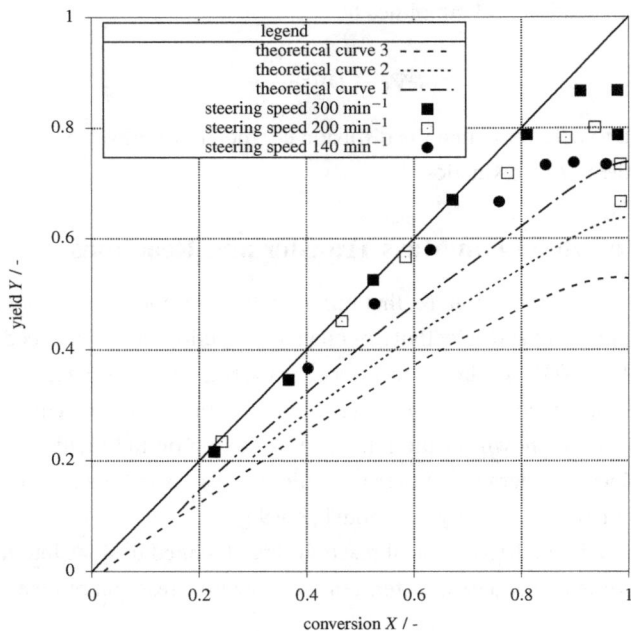

Figure 2.14 Intermediate yield for the chlorination of p-chresol in a stirred tank with constant gas-liquid contact area at different stirring speed, reprinted from [Ter69].

Teramoto et al. [Ter69] investigated the chlorination of p-cresol in a stirred tank cell regarding the intermediate yield of this gas-liquid conversion with a constant gas-liquid contact area. Figure 2.14 shows the intermediate yield in dependency of the liquid reactant conversion for three experimental investigations at different stirred speeds (300, 200 and 140 min^{-1}). It is observed, that the selectivity depends on the mixing represented by the stirrer speed. Additionally, a model based on the film theory was developed to describe this dependency. The results are also plotted within figure 2.14 and show the same trend but a not neglectable deviation to the experimental results [Ter69].

An extensive overview about the selectivity of single and two phase reactions in stirred tanks influenced by mixing is performed by Bałdyga and Bourne [Bał99, Bou03]. Besides the already mentioned influence of overall or macro mixing, determined by the mean velocity of convection, also the influence of meso and micro mixing on the selectivity is demonstrated. Meso mixing refers to the turbulent exchange on a coarse scale and micro mixing on a fine scale, where the mixing is no longer turbulent. While different experimental results and models are discussed that show an influence on yield and selectivity in case of single phase reactions, in case of two phase reactions only experimental results are available. Nevertheless, the different mixing effects also occur in gas-liquid flows. The examples for gas-liquid reactions in bulk phase, concentration film or at the interface as described in section 2.2.4 have shown, that a high selectivity is favored when the products are formed with the dissolved reactant in the film, while the side products are formed in the bulk phase [Bou03].

With the beginning 21st century, Khinast et al. [Khi01, Khi03, Koy04, Koy05] investigated numerically mass transfer and mixing sensitive consecutive reactions regarding the observed selectivity in the vicinity of gas bubbles and bubble swarms. Of special interest within these investigations is the influence of the flow profile and bubble wake structure on the selectivity. For single bubbles it is found, that in case of a rectilinear bubble rise, without wake shedding, a higher selectivity results than in case of wake shedding [Khi01]. It is also found, that the selectivity in case of a bubble swarm is lower than for a rectilinear bubble rise, but higher than in case of wake shedding at one bubble [Koy05]. With these investigations it is shown, that the influence on the selectivity through bubbly flows should to be considered for the design of gas-liquid processes. Nevertheless, experimental validation is still required to enable a reliable design of gas-liquid reactors.

Recent numerical simulations of Falcone *et. al* [Fal17] show, that the selectivity and mass transfer rate at bubbles with a size of 1.5 to 4 mm is influenced by boundary layer deformation. Furthermore, it is shown, that with rising bubble size, the dependence of the selectivity becomes less significant due to a smaller concentration boundary layer, which is less effected through the hydrodynamic conditions.

2.3 Model Reactions in Bubbly Flows

The validation of numerical simulations and the development of new measurement techniques, requires a reaction network with adjustable kinetics, that allows a detailed investigation with fluorescence techniques. Therefore, within the SPP 1740 different reaction systems are currently under development (see section 4.4.3), which are not available for first investigations. As a result, the reaction system of sodium sulfite is chosen as a model system within this work.

2.3.1 Sodium Sulfite Oxidation

The oxidation of sodium sulfite to sulfate is generally used in process engineering as a simple model reaction to determine the mass transfer coefficient. The typical reaction equation in literature is

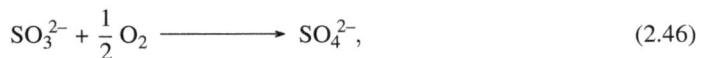

$$SO_3^{2-} + \frac{1}{2} O_2 \longrightarrow SO_4^{2-}, \tag{2.46}$$

but this formal reaction progression is only the overall reaction equation, which assembles from several partial reactions. Therefore, a very complex reaction network exists, that is not fully understood yet. Within the literature, as well as in recent publications, different reaction networks are discussed. The postulated networks differ from each other and are based on different experimental results and literature sources. Due to the so far not clarified reaction network, the reaction kinetics can not be determined without any assumptions.

The postulated reaction networks mostly differ already in the first reaction step, since the initiation is still debated. Bäckström established in 1927 [Bäc27], that the oxidation of sulfite proceeds with a radical chain mechanism and that therefore numerous radicals are involved in the reaction progression. So far the reaction step of initiation it is not clearly resolved. Furthermore, the reaction rate of the oxidation seems to depend strongly on the pH value [Lin81]. Based on these observations different reaction networks and mechanisms were postulated. Within the following the mechanism of Bäckström and the used simplified mechanism as used within this work are discussed. Additional mechanisms with a small discussion can be found in the appendix.

Reaction Mechanism according to Bäckström

In 1934 Bäckström [Bäc34] published a reaction network (figure 2.15), that is until today the basis of most mechanisms and is therefore with a few modifications generally accepted [Lin81, Sem53].

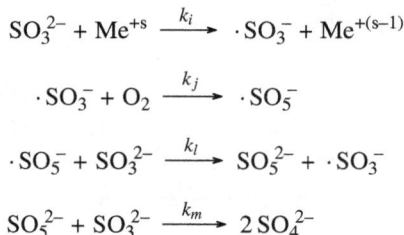

$$SO_3^{2-} + Me^{+s} \xrightarrow{k_i} \cdot SO_3^- + Me^{+(s-1)}$$

$$\cdot SO_3^- + O_2 \xrightarrow{k_j} \cdot SO_5^-$$

$$\cdot SO_5^- + SO_3^{2-} \xrightarrow{k_l} SO_5^{2-} + \cdot SO_3^-$$

$$SO_5^{2-} + SO_3^{2-} \xrightarrow{k_m} 2\,SO_4^{2-}$$

Figure 2.15 Reaction mechanism according to Bäckström [Bäc34]

This network is based on a proposal of Haber and Franck [Fra31] for the oxidation of sulfite. Bäckström explains the changes within the network on experimental findings at other radical reactions. For example Haber and Franck proposed for the last reaction step the following equation

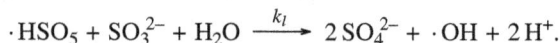

$$\cdot HSO_5 + SO_3^{2-} + H_2O \xrightarrow{k_l} 2\,SO_4^{2-} + \cdot OH + 2\,H^+.$$

But in Bäckstöms opinion, this approach is incorrect. His argumentation is based on experimental work, where alcohols are used as inhibitors in radical reactions and it is found that they result aldehydes and ketones. The alcohol is dehydrogenated, which means a loss of hydrogen atoms. Therefore it is assumed, that the sulfite ion will also lose hydrogen atoms.

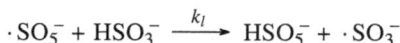

$$\cdot SO_5^- + HSO_3^- \xrightarrow{k_l} HSO_5^- + \cdot SO_3^-$$

To support the validity of this assumption, it is proposed, that during the reaction no salt of the peroxymonosulfuric acid occurs, instead the sulfate is formed. According to the properties of the two structural related substances H_2O and $K_2S_2O_8$ it can be concluded that the HSO_5^- ion reacts fast to sulfate.

Nevertheless, Bäckström did not reveal the chain termination step and progression. The available experimental data yield a lot of possible reaction and none can be isolated, but allow the assumption, that the reaction conditions influence the reaction rate of oxygen, so that no proof for the postulated mechanism is given [Bäc34]. However, Semenov [Sem53] found, that a reaction progression in the dark only occurs, if metal ions are present. Additionally, the mechanism of Bäckström does not consider the influence and dependency of the oxidation from the pH value, but in 1934 this dependency was unknown.

Simplified Mechanism

Within this work a simplified reaction mechanism according to Kück et al. (figure 2.16) is used [Küc11]. This simplification consists of less reaction steps than proposed by Ermakov und Plural or Connick *et al.* (see appendix) and is very close to the mechanism of Bäckström. It is presumed, that the reaction is catalyzed by metal ions within the solution, in case of the experimental procedure cobalt(II)sulfate.

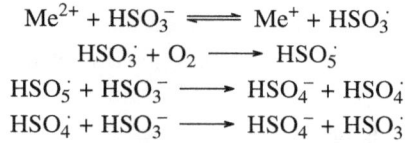

$$Me^{2+} + HSO_3^- \rightleftharpoons Me^+ + HSO_3^{\cdot}$$
$$HSO_3^{\cdot} + O_2 \longrightarrow HSO_5^{\cdot}$$
$$HSO_5^{\cdot} + HSO_3^- \longrightarrow HSO_4^- + HSO_4^{\cdot}$$
$$HSO_4^{\cdot} + HSO_3^- \longrightarrow HSO_4^- + HSO_3^{\cdot}$$

Figure 2.16 Simplified mechanism.

The simplified mechanism is also a chain mechanism, where the initiation is started by a metal ion, that is changing the oxidation state. The following step is the first chain propagating step, where a hydrogen sulfite and a hydrogen sulfate radical are formed. Afterwards an additional hydrogen sulfite radical is emerged, while chain termination steps are neglected. The advantage of this simplified mechanism is the possibility to formulate a second order kinetics easily.

$$\frac{d[O_2]}{dt} = k_1 \cdot [O_2] \cdot [HSO_3^-] \tag{2.47}$$

2.4 Conclusion

Based on the literature overview it can be concluded, that within the description of the mass transfer performance in bubbly flows and even of the mass transfer at single bubbles, still gaps and uncertainties exist. Nevertheless bubbly flows are already widely used within industrial applications, but the design and dimensioning suffers from the inaccurate description and therefore most processes can only be operated inaccurately.

An improvement in describing the mass transfer performance and a detailed understanding of the interdependency of mixing, mass transfer and chemical reaction, which is influencing yield and selectivity, is a key to lower the resource and energy consumption of most chemical and biochemical gas-liquid processes.

Chapter 3

Experimental Setups and Methods

The investigation and experimental analysis of fast reactions in bubbly flows is a challenging task due to the complex flow structures, which leads to so far unsolved relations of the time scales of mixing, mass transfer and reaction. These relations and interdependencies can't be resolved with today's techniques in a fully developed bubbly flow. To cover this gap between experimental possibilities and the requirements for a description of bubbly flows, the different timescales have to be first separated and then investigated independently to obtain a deeper insight into the processes occurring in bubbly flows.

To evaluate fast chemical reactions an experimental procedure in close collaboration with the group of Prof. Turek at the TU Dortmund is developed (see section 3.1), which allows the determination of intrinsic kinetics in a SFM. Since these investigations are performed in single phase, without mass transfer limitation, the lowest degree of complexity is assigned. Bubbly flows with a low degree of complexity are rectilinear and helical rising bubbles. Here a very reproducible trajectory and wake structure is obtained to validate numerical simulations with experimental results. By changing the boundary conditions from a free rising to a wall dominant regime, reproducible and adjustable wake structures are received. This regime is called Taylor bubble and enables detailed investigations of the influence of wake structures on chemical reactions. The highest degree of complexity investigated within this work, is the interaction of two bubbles through collision and the wobbling of free rising bubbles. An overview of the separated phenomena and the used experimental setups and methods is given in table 3.1.

Table 3.1 Separated degrees of complexity with experimental setup and methods.

Section	Phenomena	Experimental setup	Methods
3.1	Intrinsic Kinetics	SuperFocus-Mixer	Confocal Laser Scanning Microscopy
3.2	Mass Transfer through Interface and Boundary Layer - Rectilinear Rising Bubbles	Single Bubble Measurement Cell	Planar Laser Induced Fluorescence Background Illumination
3.3	Influence of Boundary Layer Deformation - Bouncing Bubbles	Single Bubble Measurement Cell	Planar Laser Induced Fluorescence Background Illumination
3.4	Influence of Mixing within the Bubble Wake	Taylor Bubble Setup	Planar Laser Induced Fluorescence Background Illumination
3.5	Influence of Boundary Layer Deformations at Free Rising Bubbles	Single Bubble Measurement Cell	Time Resolved Scanning Laser Induced Fluorescence

3.1 Intrinsic Kinetics

In case of fast chemical reactions, a mass transfer limitation in bubbly flows occurs. For the understanding, prediction and numerical simulation of chemical reactions in bubbly flows, the knowledge of the intrinsic kinetics is vital, but the determination in case of mass transfer limitation is a complex task. To eliminate mass transfer limitation, the investigations on the intrinsic kinetics have to be performed in a one phase system with a small mixing time and a high spatiotemporal resolution. To achieve concentration fields in high resolution a Confocal Laser Scanning Microscopy (CLSM) (section 3.1.2) is used in combination with a SuperFocus-Mixer (section 3.1.1) to enable time independent investigations with small mixing time.

3.1.1 SuperFocus-Mixer

Within this work a laminar interdigital micro mixer, called SuperFocus-Mixer (SFM) is used. This type of mixer (figure 3.1) is developed by Hessel in 2003 [Hes03]. Mixing in interdigital micro mixers is achieved by the use of alternating feed channels, that lead to a periodically alternating liquid lamella.

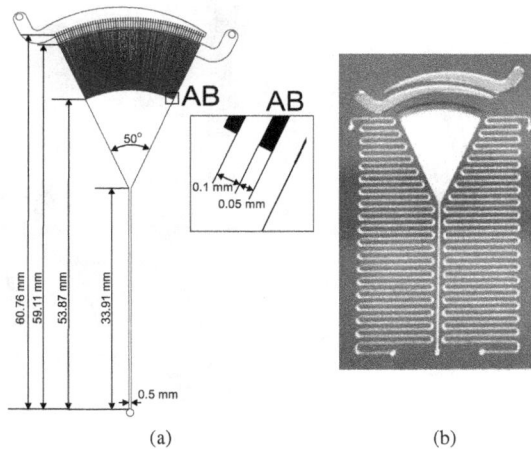

Figure 3.1 (a) Layout scheme of the SuperFocus-Mixer according to Hessel [Hes03], (b) Si-glass-SI SuperFocus-Mixer

Within the SFM 124 micro channels with a width of 100 μm are focused in a focusing chamber from a width of 20 mm to 0.5 mm to achieve fast mixing. Additionally, the inlet arrangement is curved to obtain equally distributed lamella [Hes03].

Early publications state, that the mixing time of the SFM is within a few ms [Hes03, Har03] and shortly after a mixing time of 5 ms for a 95% completion of mixing is specified [Dre04]. These mixing times seem to exclude the focusing chamber. Kashid *et. al* declared in 2014 [Kas14], that the mixing time is 4 ms for 95% completion of mixing by neglecting the residence time of the focusing chamber. The residence time is in the range of seconds to a few 100 ms depending on the flow rate, so that mixing through diffusion can not be neglected.

Nevertheless, it is shown within this work, that the laminar operating SuperFocus-Mixer is an excellent tool in combination with numerical simulations and application of the CDR-Model to determine intrinsic kinetics of chemical reactions.

3.1.2 Confocal Laser Scanning Microscopy

To obtain local concentration fields on micro-scale in high resolution, the confocal laser scanning microscopy is the technique of choice. Fig. 3.2 shows a scheme of the function principle. A CLSM works very similar to a conventional microscope. In difference, it consists additional to the lenses of two pinholes. Through these holes only focused light reaches the detector, all other light is cut off, so that a very small focus depth is reached. Therefore, only light from a chosen, narrow plane is detected and the gained information are not influenced by the planes above and below the measured plane. With deflection of the laser beam, several points in front of the microscope objective are scanned to obtain a two-dimensional plane. Every measured point is represented as a pixel in the resulting picture. The data acquisition occurs with a high resolution (below 0.35 µm per pixel), but is not instantaneous.

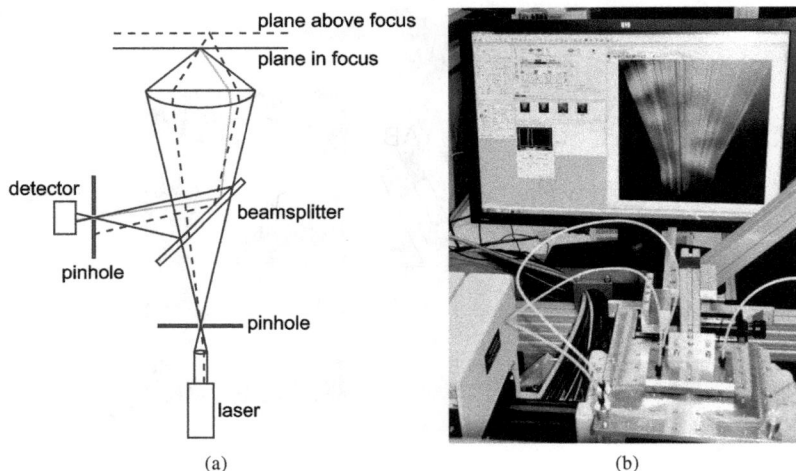

(a)   (b)

Figure 3.2 (a) Scheme of the function principle of a confocal laser scanning microscope. (b) Foto of the confocal laser scanning microscope Olympus Fluoview 1000 with attached SFM

Within this work, an Olympus Fluoview 1000 CLSM is used to measure local concentration fields within the SFM (section 3.1.1). For positioning of the mixer in front of the objective an X-Y table is developed and implemented to allow reproducible two-dimensional recordings. With change of the objective position relative to the scanned object, slices in different depth are achieved to allow a three-dimensional reconstruction of the observed concentration field.

To enable an assignment of the recorded fluorescence intensity to the concentration of an analyte, a fluorophore is necessary, that allows a reliable calibration. This is performed within the experiments with Dichlorotris(1,10-phenanthroline)ruthenium(II)hydrate (Sigma-Aldrich®), which shows a dependency of the fluorescence intensity in relation to the oxygen concentration within the solution. Through collisions of the dye molecules with oxygen, energy is transferred and the dye no longer emit energy as light. The fluorescence signal becomes lower with rising oxygen concentration. This effect is described by the Stern-Volmer correlation (see Fig. 3.3). To obtain a calibration curve, recordings of solutions with different oxygen concentrations are performed, while the concentration is monitored with a sensor (PreSens). Each measurement is performed with an aqueous solution of 30 mg·L^{-1} Dichlorotris(1,10-phenanthroline)ruthenium(II)hydrate and cobalt sulfate hydrate (16 mg·L^{-1}) in case of the catalyzed reaction. One half of the prepared dye solution is saturated with atmospheric oxygen while the other half is oxygen desorbed by the use of nitrogen. To enable reactive measurements, sodium sulfite is added to the oxygen desorbed solution.

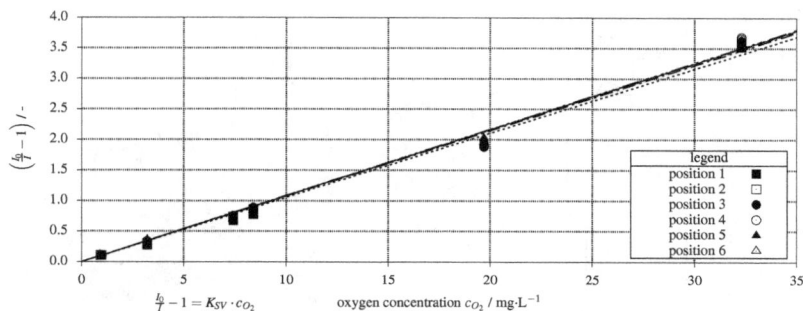

Figure 3.3 Calibration according to Stern-Volmer for the fluorescence intensity within the SFM. Data from [Spi16].

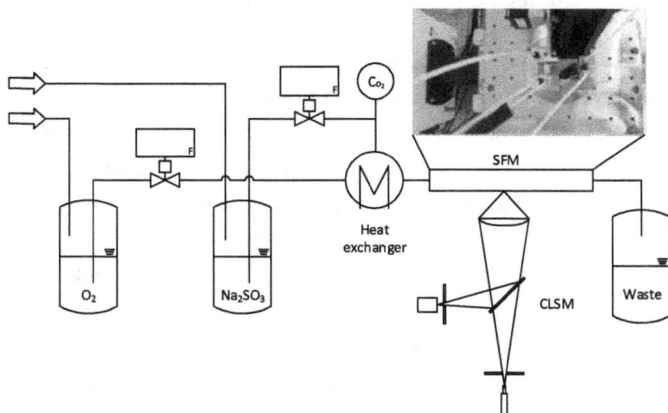

Figure 3.4 Scheme of the setup for investigations of intrinsic kinetics [Mie17].

The oxygen saturated and the sodium sulfite solution are supplied to the SFM by pressure vessels with a relative pressure of 0.5 bar. The flow rates are controlled by two Coriolis Mass Flow Meters (mini Cori-Flow M13, Bronkhorst). Preliminary investigations [Spi15] have shown a high temperature dependency of the fluorescence intensity. For this purpose, an air-conditioned box around the CLSM, an additional heat exchanger positioned in front of the SFM and a cooling zone implemented within the SFM are used to set and control the SFM temperature. With these systems a constant temperature of 20°C ± 0.5°C is ensured [Mie17].

3.2   Mass Transfer through Interface and Boundary Layer - Recti-linear Rising Bubbles

For precise investigations of mass transfer at rising bubbles as well as for the investigation in case of binary bubble interactions and dynamic shape oscillations, an easy to use and flexible basic setup is developed (see Fig. 3.5) [Tim16].

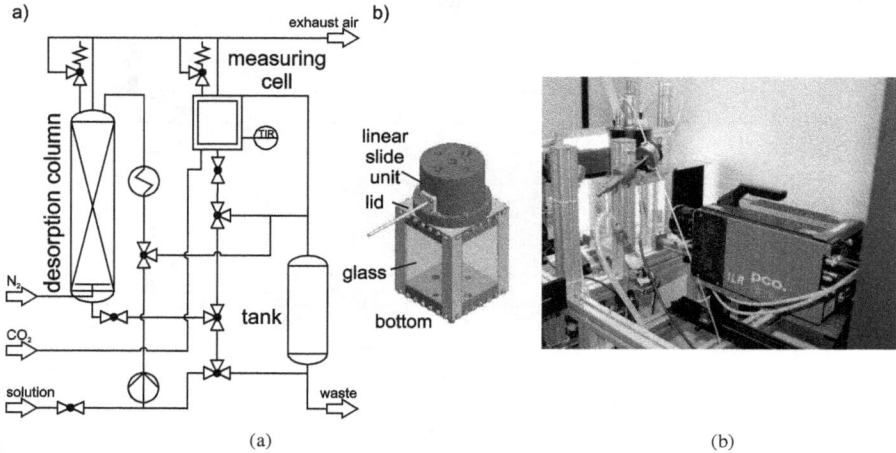

a)

exhaust air   b)

measuring
cell

linear
slide
unit

lid

glass

bottom

desorption column

N_2

CO_2

solution

tank

waste

(a)   (b)

Figure 3.5 (a) Flow scheme of the basic setup for mass transfer investigations [Tim16]. (b) Photo of the basic experimental setup.

All investigations regarding bubble rise, interaction or oscillations are performed in the single bubble measuring cell (Fig. 3.5 b). The cell has a cross section of 150 x 150 mm and a height of 200 mm. The four sides are made of two different types of glass. Perpendicular to each other two sides are made of Suprasil (UV transparent glass), while the other two sides are made of float glass (off-the-shelf glass). The two Suprasil sides are implemented to enable mass transfer investigations down to 200 nm, if necessary. The bottom and the lid of the setup are made of stainless steel. The bottom possesses six thread holes to allow different bubble generators and sensors (temperature, pressure, concentration). In the center of the lid a wide opening allows a good accessibility and is sealed during the experiments with an additional lid with attached linear slide unit to obtain oxygen exclusion.

To sharpen the measuring techniques and provide new insights into the influence of surface-active agents on the mass transfer in case of reactive bubbles, the relatively easy case of rectilinear rising bubbles according to Kück et al. [Küc09] is used. Few local mass transfer data and experimental arrangements are already available on the field of rectilinear rising bubbles [Dan07, Küc09, Jim13]. The basic setup is equipped in the center of the bottom with a bubble generation unit based on Ohl [Ohl01] to enable a generation of rectilinear rising bubbles with a size smaller than 1 mm. Two injection valves (Bosch, EV 14) are used to control the flow rates of a gaseous and liquid phase. By setting the opening duration of the two valves with a function generator, the bubble volume can be adjusted. Fig. 3.6 shows a scheme of the bubble generation unit. For a precise control of the bubble volume, the pressure is regulated by precision pressure regulators (range 0.05 to 2 bar, adjustability 0.001 bar). With this technique bubbles in the range of 0.3 to 3 mm can be generated [Ohl01]. Within this work rectilinear rising oxygen bubbles in the range of 0.5 to 0.9 mm are generated. To visualize the local mass transfer, the LIF techniques are used as described above. For this purpose the dye Dichlorotris(1,10-phenanthroline)ruthenium(II)hydrate (c = 30 mg·L^{-1}) as oxygen sensitive fluorescence marker is added to the water solution. In case of reactive mass transfer, the dye solution contains additionally cobalt sulfate (c = 16 mg·L^{-1}) as a catalyst for the sodium sulfite reaction. The camera is equipped with an Infinity K2/Distamax objective to achieve high magnifications (field of view: 7.8 x 5.9 mm).

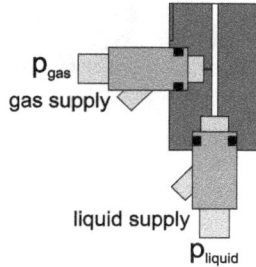

Figure 3.6 Scheme of the bubble generator.

3.2.1 Background Illumination

The determination of bubble shape, velocity and trajectory is normally performed with background illumination. Figure 3.7 shows the experimental arrangement. A light emitting diode (LED) panel powered by a rectifier LED driver is used to avoid illumination deviations of the image background at different recording speeds.

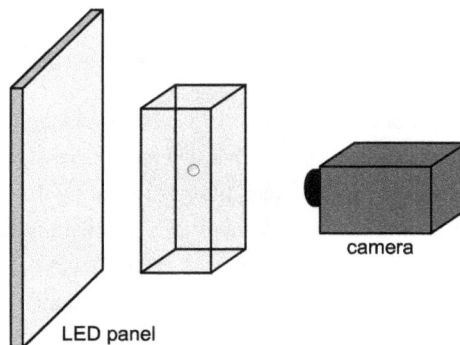

Figure 3.7 Scheme of the experimental arrangement for bubble shape, velocity and trajectory determination.

The camera (PCO Dimax HS2, PCO AG or Camrecord 500, Optronis GmbH) is positioned in view on the background illumination to record the bubble rise. An automatic determination of the bubble shape, velocity and trajectory as a MATLAB® routine is used to minimize the systematic error in case of spherical and ellipsoidal bubbles. A flowchart of the developed routine can be found within the appendix (Fig. 1). To validate the developed code recordings of a stainless steel sphere with a diameter of 5 mm is used to ensure the accurate shape and size determination. Fig. 3.8 shows an image of the stainless steel sphere and the identified sphere by the routine. The computed diameter shows an error of less than 1 %. Besides the diameter also object center position, length of the semi

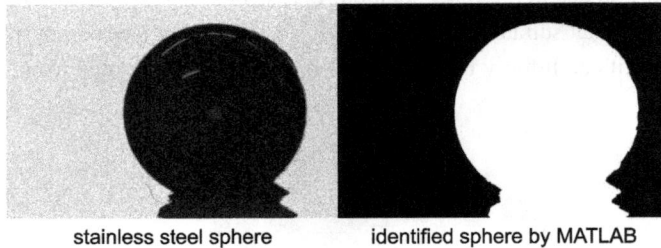

stainless steel sphere   identified sphere by MATLAB

Figure 3.8 Validation of the developed MATLAB® routine with a 5 mm stainless steel sphere.

axis and the identified area are determined. From the change of the center position on each frame and the frame rate, the object velocity is computed. In case of rising gas bubbles, this code provides the surface equivalent diameter $d_{eq,S}$, the bubble rise velocity w_b, the length of the semi axis a and b for a rotational ellipsoid and the bubble trajectory as coordinates of x and y on the recorded image.

3.2.2   Planar Laser Induced Fluorescence

While the confocal laser scanning microscopy allows a very high spatial resolution, a lack of temporal resolution has to be accepted. Within bubbly flows a high temporal resolution is necessary, since the terminal velocity observed within this work is roughly in the range of 0.1 to 0.35 m·s^{-1}, so that the bubble enters and leaves the region of interest (ROI) within a few hundred milliseconds. Therefore, the planar laser induced fluorescence (p-LIF) technique is used, which allows scan rates up to 10 000 fps.

This technique visualizes transferred oxygen similar to the described visualization in case of confocal laser scanning microscopy, but with a different periphery. Figure 3.9 shows the experimental arrangement for p-LIF investigations.

For excitation of the fluorophore a laser beam is used and widened with a light sheet optics to illuminate a planar area in the region of interest with a sheet thickness of approx. 1 mm. Perpendicular to the laser sheet, a camera, protected by a low-pass filter, is recording the emitted light. The resolution is determined by the objective in front of the camera (7 to 12 µm within the experiments). Here also a calibration procedure is required to allow a relation of the recorded gray level and is performed equal to the calibration described in section 3.1.2. For a quantitative evaluation of the recordings, in case of p-LIF, post processing of the recorded images is required. Due to a non-uniform illumination of the recorded images, a background correction is performed to obtain reliable concentration information. The correction procedure is based on the work of Dani et al. [Dan07]. First of all, an image sequence of oxygen desorbed solution, as background information, is recorded and an averaged image using MATLAB® is computed. The raw images of the bubble rise (figure 3.10 a) are divided by the averaged background image and are afterwards normalized, so that a nearly uniform background is obtained (see figure 3.10 b). Since high speed recordings are performed, the background noise is decreased

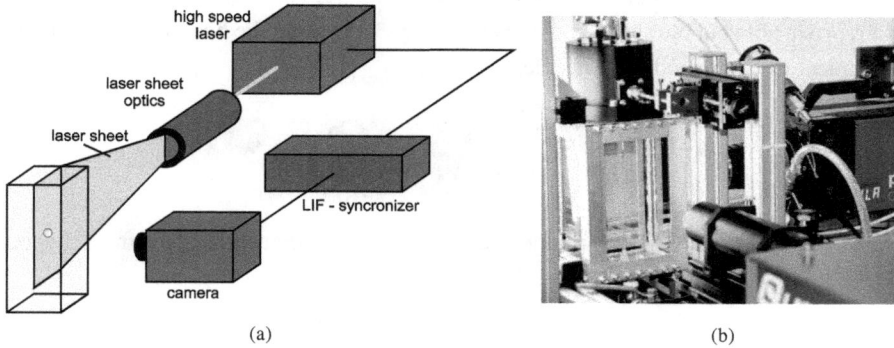

Figure 3.9 (a) Scheme of the experimental setup for p-LIF measurements. (b) Photo of the experimental setup for p-LIF measurements.

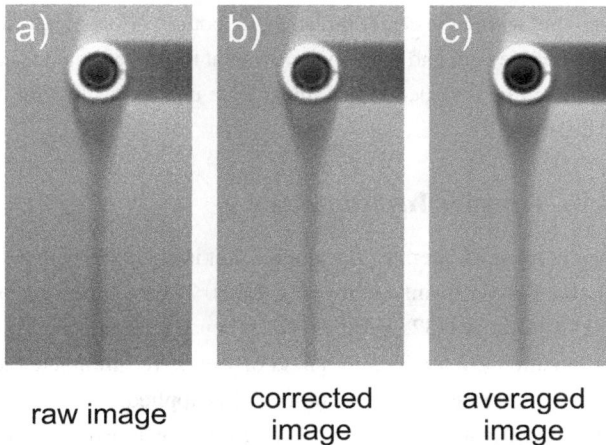

Figure 3.10 Image processing of raw images.

by a special averaging of the recording sequence. Through an automatic processing of the images using edge detection, MATLAB® specifies the center of the bubble on each image. Afterwards the bubble center is used to adjust the region of interest frame (400 x 800 pixel), so that the bubble center is always on the position of 200, 150 pixels. The resulting ROI frames are averaged to obtain one image with small noise information (see figure 3.10 c).

3.3 Influence of Boundary Layer Deformation - Bouncing Bubbles

Binary bubble interactions occur typical stochastically due to the complex flow structures in bubbly flows. Nevertheless, accurate investigations require a high degree of reproducibility. Therefore, an experimental procedure is developed that allows generation of bubbles with reproducible shapes and trajectories. A hypodermic needle is used as orifice according to Sone et.al. [Son08]. At this orifice the bubble interface is always deformed at the same position for each equivalent bubble volume and leads to a reproducible bubble shape and trajectory. Fig. 3.11 shows the boundary layer at and close after detachment from a hypodermic needle.

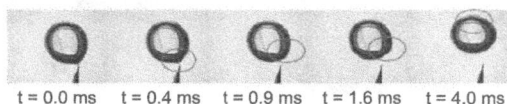

t = 0.0 ms   t = 0.4 ms   t = 0.9 ms   t = 1.6 ms   t = 4.0 ms

Figure 3.11 Boundary layer deformation close and after detachment from a hypodermic needle [Tim16].

To control the bubble volume here as well injection valves are used. In difference to the already described bubble generator no water is supplied and only the flow rate of the gaseous phase is controlled. Through combination of these two techniques, a very reproducible bubble generation becomes possible. The experimental arrangement is already shown in Fig. 3.5 a). The hypodermic needle is positioned rotatably within the center thread in the bottom of the measuring cell, to enable the precise adjustment of the bubble path and collision position at the upper bubble. For the positioning of the upper bubble, the lid opening is sealed with a linear slide unit to match with the bubble trajectory of the rising bubble [Tim16].

3.3.1 Overall Mass Transfer Investigations

For determining the overall mass transfer, the linear slide unit is equipped with a device to fix the CO_2 bubble at the top to enable bubble shrinkage investigations. The volume decrease is recorded with a high speed camera (PCO Dimax HS2, 5 frames per second) with attached Infinity K2/Distamax objective for high magnification (52.5 x 39.5 mm field of view). To illuminate the background of the cell a LED panel as already described within section 3.2.1 is applied.

For the experimental investigations, the setup is filled with deionized water. As gases CO_2 and N_2 are used. The rising and the mechanically fixed CO_2 bubbles are generated with a size of 1.72 $\pm$ 0.05 mm. To investigate the influence of collision frequency (0, 0.05 and 2 s^{-1}) on mass transfer the time between bubble generation (0, 0.05 and 20 s) is varied within the experimental procedure [Tim16].

3.3.2 Local Mass Transfer Investigations

To get a deep insight in the mechanism of mass transfer enhancement in case of bubble bouncing, the local oxygen concentration field is investigated by p-LIF (see section 3.2.2). Within these investigations the fluorophore Dichlorotris(1,10-phenanthroline)ruthenium(II)hydrate with a concentration of 30 mg·L^{-1} is as well used. The dye is excited with a Nd:YLF laser (wavelength 527 nm, pulse width < 210 ns, pulse repetition rate 1 kHz, Continuum®). The fluorescence light is recorded perpendicular to the laser light sheet by a PCO Dimax HS2 (1000fps). To protect the camera from direct laser radiation, a bandpass filter (center wave length 590 nm $\pm$ 2 nm, half-power bandwidth 20 nm $\pm$ 2 nm, transmission > 84%, ILA_5150 GmbH) is used.

In difference to the investigations on the overall mass transfer, a capillary (inner diameter: 0.2 mm) is used at the position of fixation device to generate an upper N_2 bubble to visualize the concentration field of the rising O_2 bubble only. The capillary is also connected, like the hypodermic needle within the bottom, to an injection valve for the generation of reproducible bubbles and is mounted on the linear slide unit. A delay of 100 ms between generation of the fixed upper and rising bubble is used to ensure an uninfluenced bubble generation of the lower, rising bubble. The camera recording is controlled by a light barrier to allow a better utilization of the data storage, so that many bubble collisions are observed within one recording. Figure 3.12 shows a scheme of the used setup.

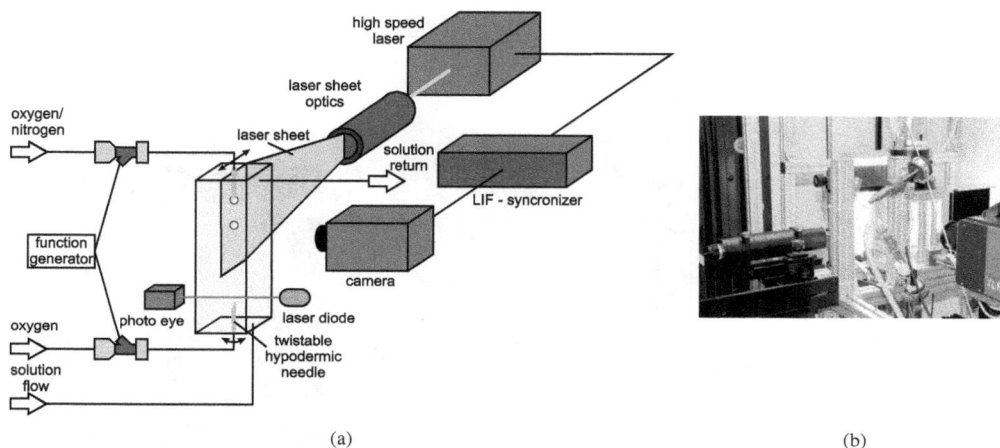

(a)   (b)

Figure 3.12 (a) Scheme of the experimental setup for the investigation of local mass transfer using p-LIF [Tim16]. (b) Photo of the experimental setup for the investigation of local mass transfer using p-LIF.

Beside the influence of bouncing on physical mass transfer, also the influence on reactive mass transfer is investigated. For this purpose, defined conditions, under N_2 atmosphere, are realized within the measuring cell by flushing the cell with nitrogen for 30 minutes. Then the fluorophore solution, with additionally cobalt(II)sulfate as a catalyst for the reaction, is filled into the desorption column and is saturated with nitrogen. Afterwards the oxygen desorbed solution is filled into the measuring cell and the oxygen concentration is logged with a sensor probe (PreSens) during the whole experiment.

The reaction is a catalyzed oxidation of sodium sulfite according to figure 2.16. While the overall reaction equation is simple, the reaction network is very complex due to the radical character (see section 2.3.1). Nevertheless, the timescales of this reaction can be adjusted very easily over a wide range and is therefore used commonly for the determination of mass transfer coefficient and is used within this work for a first investigation on the influence of bubble bouncing on chemical reactions [Tim16].

3.4 Influence of Mixing within the Bubble Wake - Taylor Bubble

To identify suitable reaction systems and adjust the timescales of chemical reaction to the timescales of mixing and mass transfer, an experimental setup is required, that allows a reproducible mass transfer and flow pattern. Taylor bubbles, air bubbles bigger than the inner diameter of the capillary, rise with a constant velocity over a wide range of bubble equivalent diameter, so that quasi steady state investigations can be performed.

Figure 3.13 shows the experimental setup, which consists of a glass channel with an upper storage tank and is developed based on the guiding measure of the SPP 1506 "Transport Processes at Fluidic Interfaces".

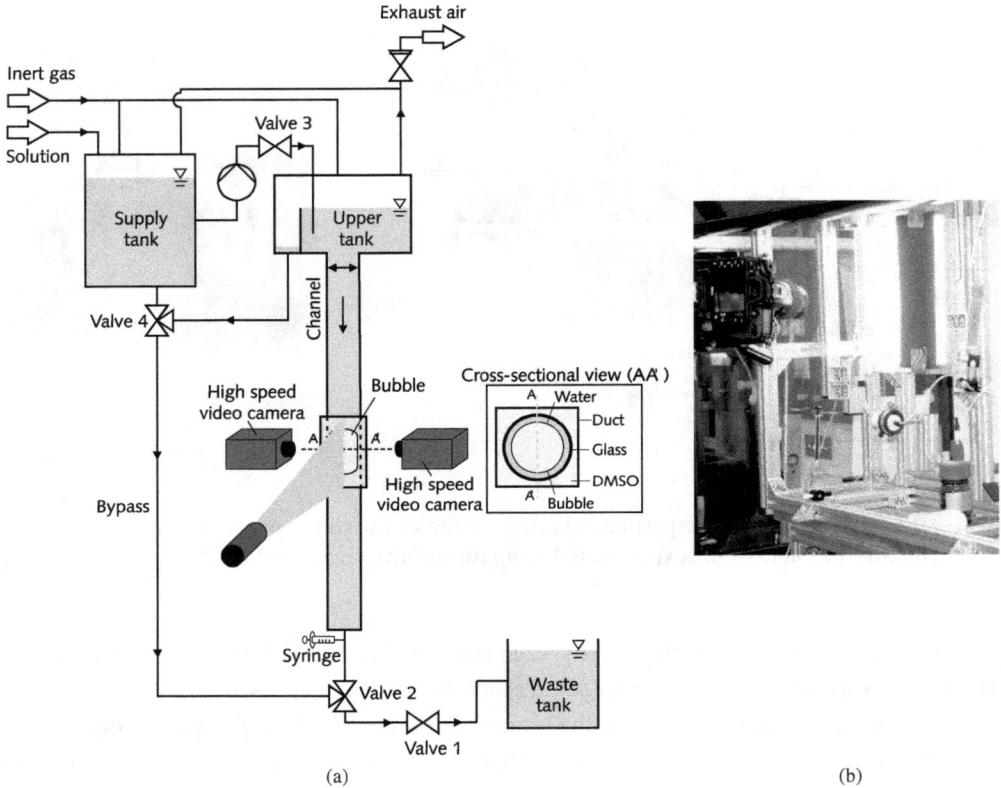

Figure 3.13 (a) Scheme of the setup for investigations of Taylor bubbles [Kas17]. (b) Photo of the experimental setup for investigations of Taylor bubbles.

The flow rate within the glass channel (length 300 mm) is controlled by valve 1 to fix the bubble hydro-mechanically at a position of 360 mm below the liquid surface in the upper tank. The gas is injected by the use of a gas-tight syringe (Hamilton 1001) into the lower part of the channel by using a 3-port valve. To adjust the rise velocities and wake structures in case of different fluid properties, different circular test sections (diameter: 4 to 8 mm; Glastechnik Kirste KG) are used. For distortion-free recording, the channel is placed inside of a cuvette which is filled with a solution of water and dimethyl sulfoxide (DMSO, Sigma Aldrich) of 97 wt% at 298 ± 1.0 K for refractive index matching. The entire setup is flushed for 30 minutes before each experiment with nitrogen to ensure oxygen exclusion [Kas17].

3.5 Influence of Boundary Layer Deformations at Free Rising Bubbles

With increasing complexity and rising bubble equivalent diameter, the investigated partial processes are no longer reproducible. A high degree of reproducibility is necessary for p-LIF investigations since the light sheet position can not be altered easily during the measurement. Due to the statistical bubble rise and occurring boundary layer deformations in this regime, a three-dimensional technique is required to reveal the mass transfer processes.

3.5.1 Time Resolved Scanning Laser Induced Fluorescence - TRS-LIF

Within this work a time resolved scanning LIF (TRS-LIF) technique is used and further developed to visualize concentrations fields and wake structures [Brü99, Deu01, Cri08, Stö09, Soo12]. Fig. 3.14 shows a scheme of the experimental arrangement, which is based on the work of Soodt et al. [Soo12].

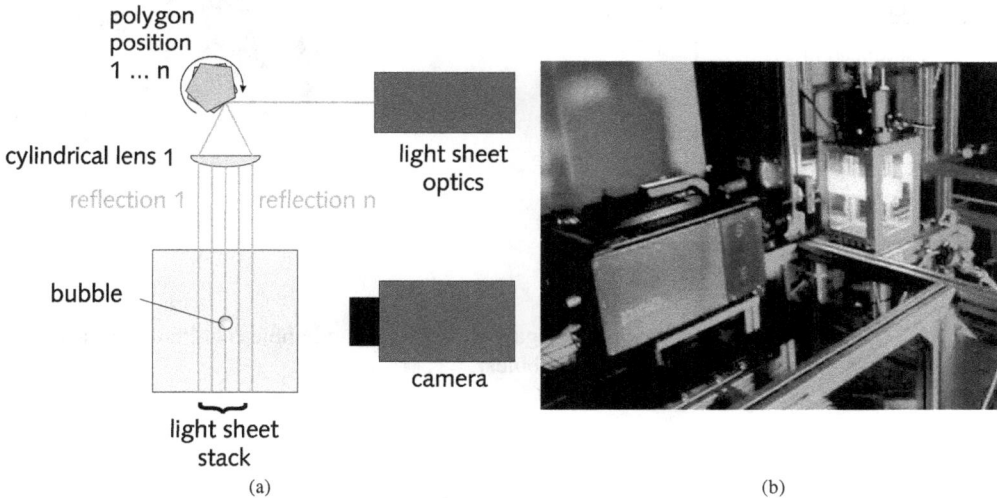

Figure 3.14 (a) Experimental setup for TRS-LIF investigations based on Soodt *et al.* [Soo12]. (b) Photo of the experimental setup for TRS-LIF investigations with laser emission.

In difference to p-LIF, a rotating polygon reflects the laser beam into the volume of interest. Through a precise synchronization of rotation and laser emission, a reproducible specular deflection results. Afterwards the light sheet is parallelized by a cylindrical lens to obtain equidistant spacing. The number of light sheets in a stack is determined by the mirrors per polygon, the laser repetition rate, the polygon dimensions and the rotational speed [Soo12]. With the used polygon (20 mirrors, 75 mm in diameter), a rotational speed between 1000 to 10000 rpm and a laser repetition rate up to 20 kHz, 1 to 60 slices with a minimum spacing of 0.9 mm can be obtained.

3.5.2 Dynamic Shape Oscillations at Free Rising Bubbles

The highest degree of complexity investigated within this work are dynamic shape oscillations at free rising oscillating bubbles. The three-dimensional concentration wake is reconstructed based on the investigations with TRS-LIF (see 3.5.1). Therefore, the measuring cell is equipped with an orifice (inner diameter 3 mm) attached to the bubble generator according to Ohl [Ohl01] and filled with nitrogen saturated fluorophore solution. Bubbles in the range of 4 to 8 mm are generated every 20 seconds to observe dynamic shape deformations in a stagnant liquid.

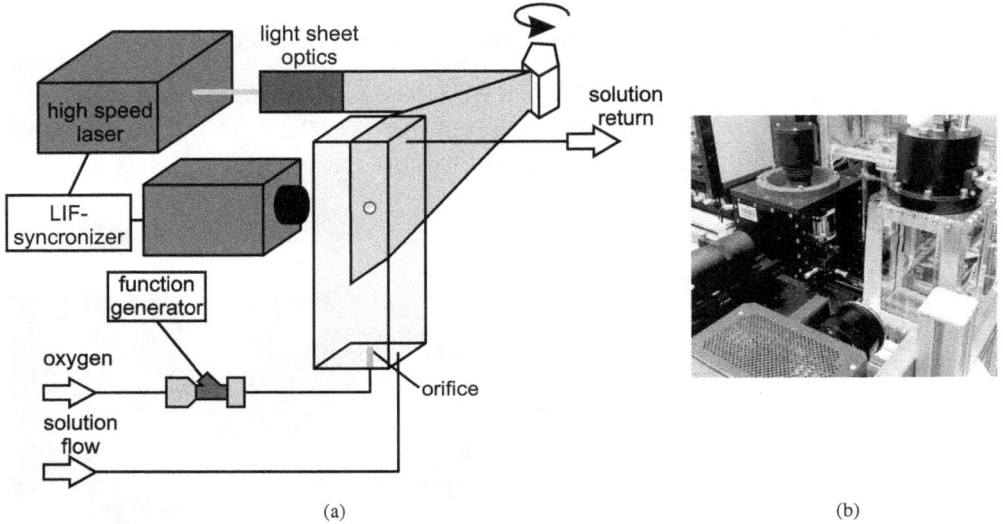

Figure 3.15 (a) Scheme of the setup for investigations at free rising bubbles (b) Photo of the experimental setup for investigations at free rising bubbles.

Chapter 4

Experimental Results and Discussion

To clarify the influence of the time scales of mixing and mass transfer on chemical reaction, the five different experimental setups according to table 3.1 have been used. The results are structured according to their level of complexity.

4.1   Intrinsic Kinetics

The determination of intrinsic kinetics is a key task to clarify the interdependence of timescales of mixing, mass transfer and chemical reactions. Especially for numerical simulations, the knowledge of the intrinsic kinetics is crucial, since reliable prediction of chemical processes is only possible with exact values.

The evaluation of kinetics without mass transfer or diffusion limitation is a challenging task due to mixing through diffusion, which occurs even on the smallest scale within a given time. To eliminate the influence of mass transfer, the usage of micro mixers in single phase is the technique of choice. Since diffusion and convection could not be neglected within micro fluidic devices, a direct evaluation of intrinsic constants is impossible. Therefore, in close collaboration with the group of Prof. Turek at the TU Dortmund, a method for the evaluation, based on the software FEATFLOW, for the intrinsic kinetics is developed.

4.1.1 Comparison of Physical Mixing with Numerical Simulations

For the comparison of the experimental and numerical concentration fields within the SFM nine different positions are investigated according to (Figure 4.1).

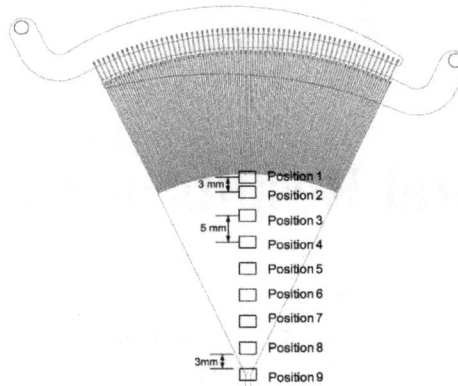

Figure 4.1 Scheme of the investigated positions within the mixing chamber of the SFM. Adapted from [Spi16]

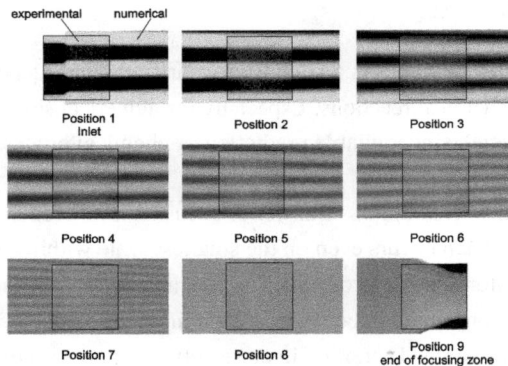

Figure 4.2 Experimental concentration fields embedded within the numerical results within the SFM in chase of physical mixing at 150 g·h^{-1} [Mie17]

To proof the comparability of experimental investigations and numerical simulations, the SFM is supplied with a nitrogen ($c_{O_2} = 0$ mg·L^{-1}) and an oxygen saturated ($c_{O_2} = 40$ mg·L^{-1}) fluorophore solution with a flow rate of 30, 150 and 300 g·h^{-1}. These parameters are applied as initial conditions for the numerical simulation with the diffusion coefficient $D_{O_2} = 2.0 \cdot 10^{-9}$ m^2·s^{-1}. Figure 4.2 shows exemplarily the recorded concentration fields embedded within the numerical results from the group of Prof. Turek. An satisfactory agreement is achieved, while nevertheless small deviations are observed due to defects within the geometry of the SFM.

As a further validation, the influence of different depth is investigated. Closer to the mixer wall, diffusion is pronounced, since the velocity of the fluid is lower, due to the no slip condition. Figure 4.3 shows the obtained results for position 1, 2 and 3 in a depth of 0.15, 0.08 and 0.02 mm. Here an excellent agreement between simulation and experiment for diffusion and convection is obtained. Based on these agreements it is expected, that through the application of numerical simulation in cooperation with experimental results the clarification of intrinsic kinetics with superimposed diffusion and convection are achieved [Spi16].

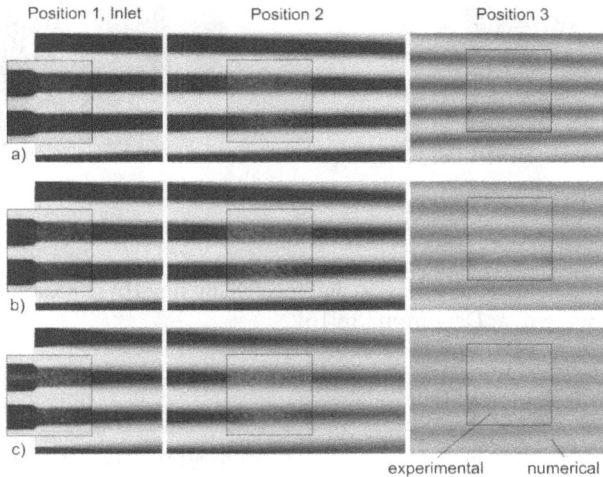

Figure 4.3 Comparison of experimental and numerical concentration fields within the SFM in case of physical mixing at 150 g·h^{-1} in the depth of a) 0.15 mm, b) 0.08 mm and c) 0.02 mm [Mie17]

4.1.2   Comparison of Physical and Reactive Mixing

As a first attempt to evaluate intrinsic reaction constants, the catalyzed oxidation of sodium sulfite is used (section 2.3.1), since high sodium sulfite concentration in the uncatalyzed case already influence the liquid properties. For the determination of the intrinsic kinetics, the reaction parameters have to be adjusted. Therefore, the oxygen concentration is lowered to 12 mg·L^{-1} (3.75·10^{-4} mol·L^{-1}), while the sodium sulfite concentration is set to 0.4 g·L^{-1} (3.17·10^{-3} mol·L^{-1}), so that an sulfite excess of approximately 10 with respect to oxygen results. Figure 4.4 shows in comparison the concentration fields in case of physical and reactive mixing at a flow rate of 300 g·h^{-1}. Due to the superimposed sodium sulfite reaction, the oxygen enriched stream (red) is narrowing much faster. Additionally, the diffusion zone, as in the case of physical mixing, is nearly completely consumed by the reaction.

Figure 4.5 shows mean concentration plots of the pixel lines 295 to 305 from the concentration fields in Figure 4.4. It is notably, that the maximal and minimal fluorescence intensity is lower in case of reactive mixing, which leads to a shift in the evaluated concentration, due to a change of the pH value of the sodium sulfite solution. Nevertheless, the influence of the catalyzed reaction is clearly observable. On one hand, the concentration gradient is slightly steeper within the reactive profile. On the other hand, the ratio of oxygen enriched to an depleted area is shifted to an expanded depleted area, as expected.

To obtain additional validation data, also the flow rates of 30 and 150 g·h^{-1} are investigated. By comparing the oxygen concentration fields at 150 g·h^{-1} (see Figure 4.6), the reaction influence becomes more clear. The thinning of the oxygen enriched areas (red) is more distinct, while the

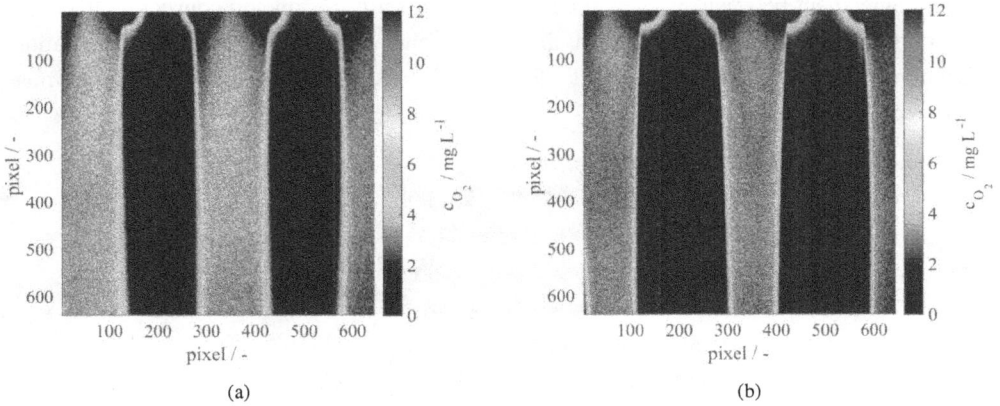

(a)   (b)

Figure 4.4 Concentration fields of physical (a) and reactive mixing (b) at position 1 (inlet) within the SFM at a flow rate of 300 g·h^{-1}. Data from [Spi16].

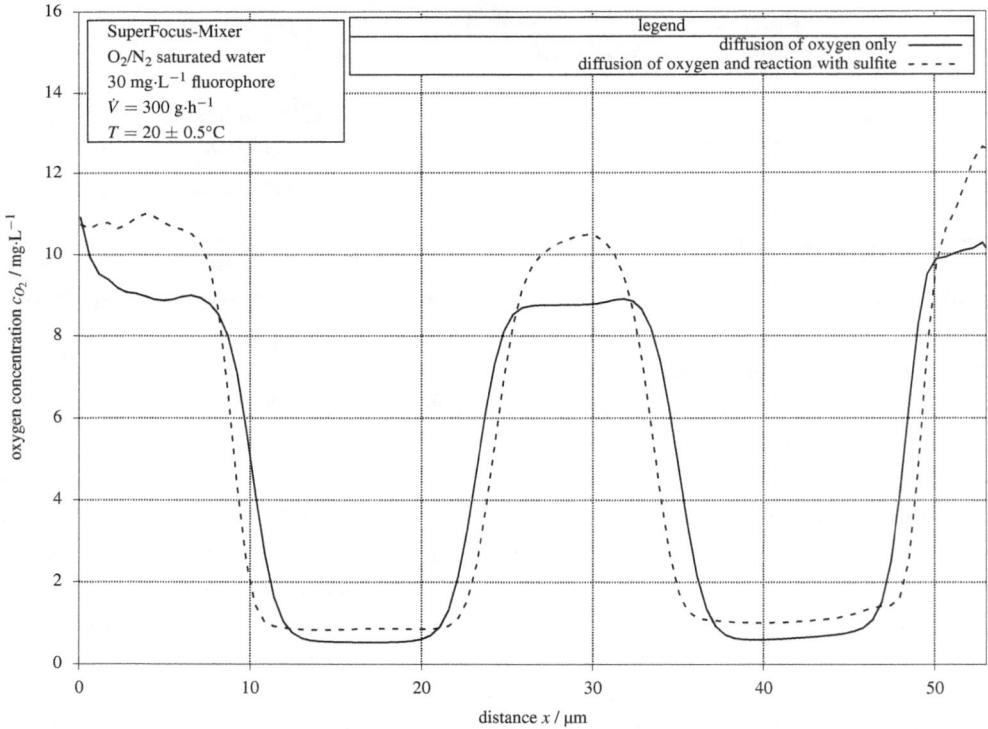

Figure 4.5 Mean concentration plots for 10 scan lines of physical (black) and reactive mixing (dashed) at position 1 (inlet) within the SFM at a flow rate of 300 g·h^{-1}. Data from [Spi16].

oxygen depleted areas (blue) are much broader in the case of reactive mixing (see b). It is observed in case of physical mixing, that the diffusion is more pronounced according to the theory, since the volume flow is 50% smaller and therefore the residence time is two times longer.

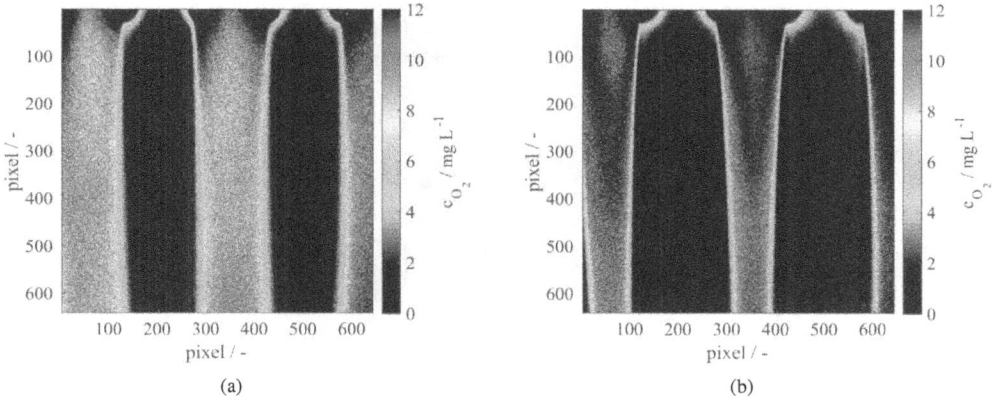

Figure 4.6 Concentration fields of physical (a) and reactive mixing (b) at position 1 (inlet) within the SFM at a flow rate of 150 g·h^{-1}. Data from [Spi16].

The concentration plots (figure 4.7) show the same behavior than the concentration fields. The gradient between low and high concentration, in case of physical mixing, is broader as for the higher volume flow, while it is still steep for reactive mixing.

With a further decrease of the flow rate, down to 30 g·h^{-1}, the diffusion is again more pronounced, so that in case of physical mixing a broad diffusion zone on the recorded 533 µm results (Figure 4.8). In case of reactive mixing, nearly all oxygen is consumed by the reaction after approximately 300 µm due to the 10 times higher residence time than in figure 4.4.

Figure 4.9 also illustrates these findings. The diffusion zones are not as steep as in the two cases before and the oxygen concentration is lower in case of physical mixing in comparison to the other two flow rates, while in case of the reactive mixing, the oxygen is almost consumed at this position.

4.1.3   Determination of the Intrinsic Kinetics

As already mentioned within the introduction of this section, the determination of intrinsic kinetics in micro reactors is performed by the use of experimental results and validation of numerical simulations with a parameter optimization. Besides the material properties, like viscosity, concentration and geometry, the diffusion coefficients of the different species are required. The coefficients for sulfite and sulfate are equal with a value of $D_{SO_3^{2-},SO_4^{2-}} = 1.0 \cdot 10^{-9}$ m·s^{-1} [Lea85], while for oxygen a value of $D_{O_2} = 2.0 \cdot 10^{-9}$ m·s^{-1} is used [Dav57]. Additionally, the influence of fluorophore and catalyst is neglected. Since the numerical simulation can't handle different diffusion coefficients, one value for an effective diffusion D_{eff} is implemented in the optimization process, so that a two parameter optimization results.

With the concentration fields at given axial positions within the SFM center and the three different flow rates of 300, 150 and 30 g·h^{-1} as constraints, the optimization is performed at the TU Dortmund. Variation of the flow rate is necessary to cover the full performance of the SFM. At the highest flow

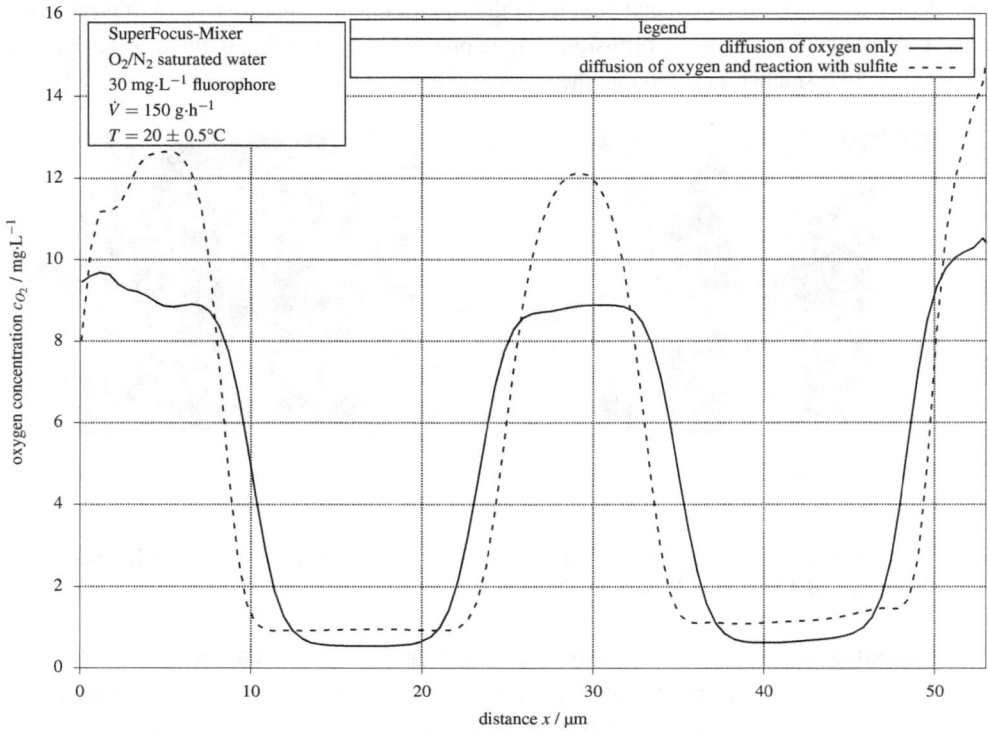

Figure 4.7 Mean concentration plots for 10 scan lines of physical (black) and reactive mixing (dashed) at position 1 (inlet) within the SFM at a flow rate of 150 g·h^{-1}. Data from [Spi16].

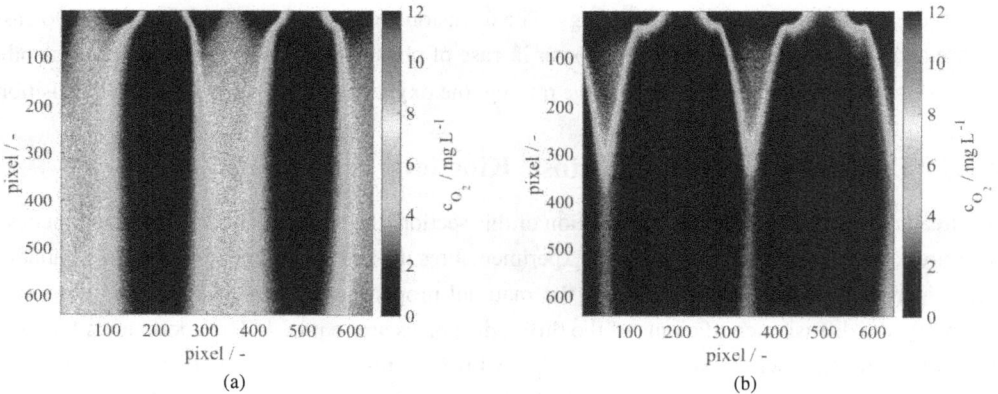

Figure 4.8 Concentration fields of physical (a) and reactive mixing (b) at position 1 within the SFM at a flow rate of 30 g·h^{-1}. Data from [Spi16].

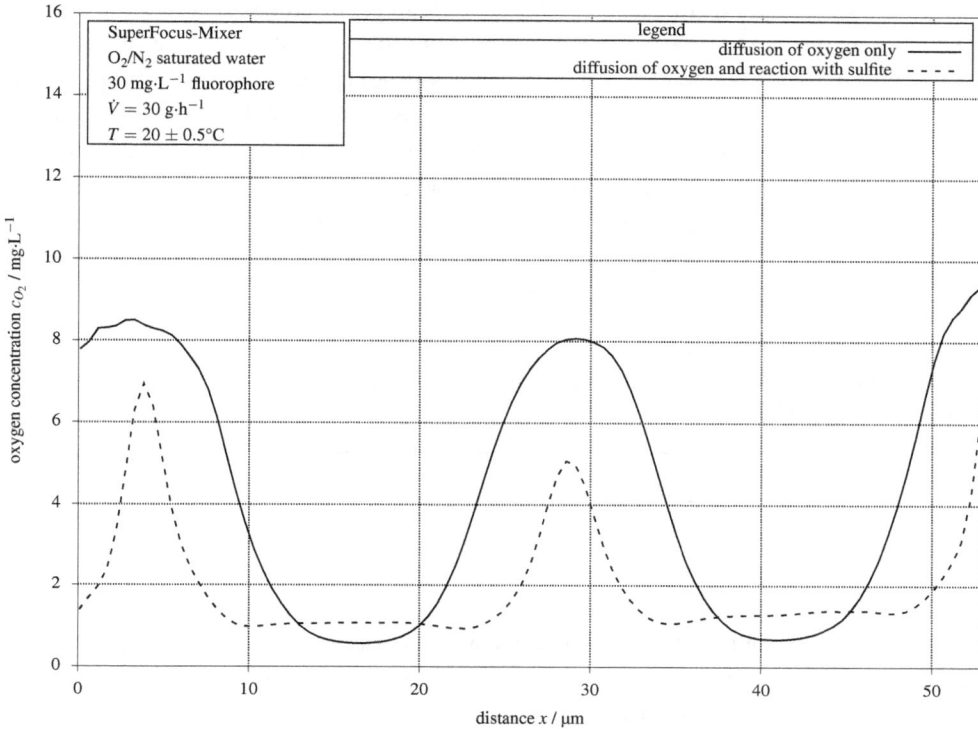

Figure 4.9 Mean concentration plots for 10 scan lines of physical (black) and reactive mixing (dashed) at position 1 within the SFM at a flow rate of 30 g·h^{-1}. Data from [Spi16].

rate, the residence time is relatively low, so that mixing through diffusion, enhanced by the reaction, is insufficient to homogenize the streams. By lowering the volume flow, the diffusion is more and more pronounced, so that the oxygen enriched stream is consumed shortly behind the inlet.

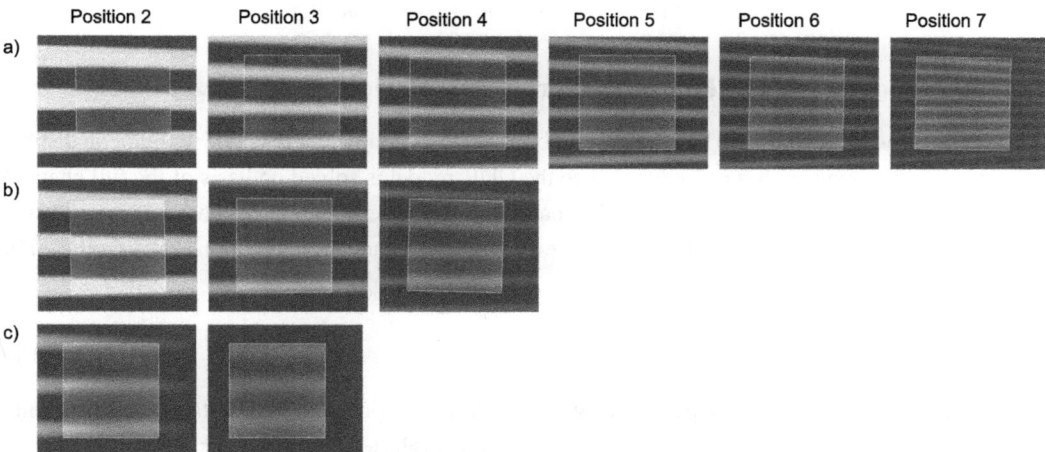

Figure 4.10 Simulated concentration fields with embedded experimental results for a flow rate of (a) 300, (b) 150 and (c) 30 g·h^{-1} [Mie17].

Figure 4.10 shows the experimental results embedded in the recording positions within the simulated concentration fields. The parameters with best fit and error to the experimental results are summarized in table 4.1. According to the work of Kück [Küc11], the estimated reaction rates for sodium sulfite in this concentration range are between $k_{CR2} = 1 \cdot 10^4$ to $100 \cdot 10^4$ L·mol^{-1}·s^{-1}, so that the determined reaction rate is in a good agreement. Nevertheless, there are still deviations between experimental and numerical results. It is observed, that for regions with a high oxygen concentration (above 2 mg·L^{-1}) a very good agreement is accomplished, while for lower oxygen concentrations a mismatch results. This finding is probably based on a change of reaction network as observed by Linek *et al.* [Lin81], due to the low oxygen concentration.

Table 4.1 Listing of diffusion coefficient and reaction rate, determined for the reaction of O$_2$ with sodium sulfite in a SFM [Mie17]

result	value	unit
D$_{eff}$	$1.0 \cdot 10^{-9}$	m^2·s^{-1}
k$_{CR2}$	$5.5 \cdot 10^4$	L·mol^{-1}·s^{-1}

The combination of numerical and experimental investigations within the SFM have proven to be an excellent tool for the determination of intrinsic reaction parameters and can be easily adapted to any reaction system if one of the species, in the best case the product, is observable with confocal laser scanning microscopy. Through the conversion of the reaction time scale on a length scale in steady state operated mircofluidic devices, a time independent investigation is possible with low reactant consumption. Nevertheless, the mixing through diffusion is not neglectable in a lamination device as the SFM, so that diffusion has to be taken into account for the determination of the time constants. It is likely, that also parallel consecutive reactions can be investigated with this technique to determine the influence of mixing time on the yield and selectivity in a single phase easily. Within the following, the obtained kinetics are used for further investigations in two phase flow operations.

4.1.4   Modeling of Reactions in Microfluidic Devices

The evaluation of the intrinsic kinetics by combination of experimental results with numerical simulations is a time-consuming procedure, since the CDR-model is solved on basis of the full channel geometry without any simplifications. To enable a faster determination of the kinetics, a simplified CDR-model is developed. Therefore, a few assumptions and simplifications are introduced in equation 4.1

$$v\nabla c = D\nabla^2 c \pm r_k \ . \tag{4.1}$$

Since the measurements are performed in a microfluidic device with steady state conditions, only a one dimensional problem results, if the channel depth is neglected [War10]. Therefore Fick's second law (equation 4.2)

$$\frac{\partial c}{\partial t} = D\frac{\partial^2 c}{\partial x^2} \tag{4.2}$$

can be solved analytically with the following start conditions for concentration A and B

$$c_{A,t=0} = c_A (x,t = 0) = c_{A,0} \ if \ x \leq 0 \tag{4.3}$$

$$c_{B,t=0} = c_B (x,t = 0) = 0 \ if \ x \leq 0$$

$$c_{A,t=0} = c_A (x,t = 0) = 0 \ if \ x > 0 \tag{4.4}$$

$$c_{B,t=0} = c_B (x,t = 0) = c_{B,0} \ if \ x > 0 \ .$$

With these conditions, the analytical solution of equation 2.34 is the error function *erfc* [Cra79]

$$c(x,t) = \frac{1}{2} \cdot c_0 \cdot erfc \frac{x}{2\sqrt{D \cdot t}} \tag{4.5}$$

Due to the steady state within the SFM, the reaction time is transformed to a spatial coordinate and the development of concentration for reactant A, $c_{A,x,y}$ is described by

$$c_{A,x,y} = \frac{1}{2} c_{A,0} \ erfc \frac{y}{2\sqrt{D_A \cdot x \cdot v^{-1}}} - c_{P,x,y-1} \tag{4.6}$$

with $c_{A,0}$ as the start concentration of A, the spatial coordinates x and y, D_A the diffusion coefficient of species A and the product concentration $c_{P,x,y-1}$ of the former time step. It have to be noted, that the product concentration at the position $c_{P,x,0}$ is set as zero to fulfill the boundary conditions.
For the concentration of reactant B ($c_{B,x,y}$), which diffuses from the opposite site of the SFM the diffusion zone is described by

$$c_{B,x,y} = c_{B,0} \left(1 - \frac{1}{2} \ erfc \frac{y}{2\sqrt{D_B \cdot x \cdot v^{-1}}} \right) - c_{P,x,y-1} \tag{4.7}$$

with $c_{B,0}$ as the start concentration of B and D_B the diffusion coefficient of species B.
The concentration field of the product is evaluated, by solving the differential equation for a reaction of second order with unequal start concentrations

$$\frac{dc}{dt} = -k \cdot c_A \cdot c_B \ . \tag{4.8}$$

For solving this equation, at first the concentrations of c_A and c_B are substituted by $c_{A,0} - c_P$ and $c_{B,0} - c_P$ to separate the variables. Afterwards a partial fraction expansion with following integration is performed, so that

$$\ln \frac{c_{A,0} - c_P}{c_{B,0} - c_P} = (c_{A,0} - c_{B,0})kt + \ln \frac{c_{A,0}}{c_{B,0}} \tag{4.9}$$

results. For the evaluation of the product concentration field, equation 4.9 is transposed to c_P and the concentration $c_{P,x,y-1}$ of the former time step is added

$$c_{P,x,y} = \frac{c_{A,x,y} \cdot c_{B,x,y}[exp\{-k\Delta t \cdot (c_{A,x,y} - c_{B,x,y})\} - 1]}{c_{B,x,y} \cdot exp\{-k\Delta t \cdot (c_{A,x,y} - c_{B,x,y})\} - c_{A,x,y}} + c_{P,x,y-1} \; . \tag{4.10}$$

The time difference Δt is described by the resolution of the computational grid z_{GRID} divided by the flow velocity $\Delta t = z_{GRID}/v$. With equation 4.6, 4.7 and 4.10 the concentration fields of the reactants A (O_2) and B (SO_3^{2-}) and the product P (SO_4^{2-}) can be computed. For this purpose, the diffusion coefficients and reaction constants as applied and found by Mierka [Mie17] (section 4.1.3, see table 4.1) are used. In case of physical mixing, without the sodium sulfite reaction, only equation 4.6 is solved. As a simplification the wall with a size of 20 μm at the beginning of the mixing chamber is neglected and additionally only one mixing field with a size of 120 x 533 μm is computed (see figure 4.11).

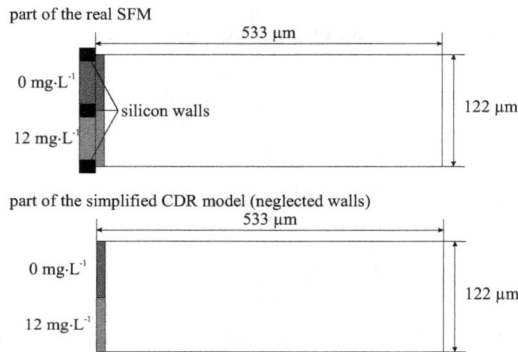

Figure 4.11 ROI in case of experimental and calculated concentration field.

For a direct comparison mean concentration plots of the grid lines 295 to 305 are computed similar to the experimental plots. Figure 4.12 shows the comparison of the CDR model and the experimental results (see also figure 4.5, 4.7 and 4.9).

By comparing the computed concentration profiles for physical and reactive mixing at the highest flow rate of 300 g·h^{-1} (figure 4.12 a) with the measured profiles, a good agreement of the flank position and gradient is reached, which justifies the findings within section 4.1.3.

For a further validation the flow rates of 150 and 30 g·h^{-1} are evaluated. For a flow rate of 150 g·h^{-1} it is found, that the position of the flanks and the gradient are very similar between the CDR-model and the experimental results. Nevertheless, the absolute oxygen value of the experimental findings is different from the idealized model as well. In case of the flow rate of 30 g·h^{-1} there is a high mismatch between the position of the flank in case of physical mixing, while the gradient is very similar. Additionally, a high match of the position and gradient in case of reactive mixing is found. The mismatch is very likely a result of the high error of the mass flow controllers at the bottom range of adjustment (see also section 4.6). Additionally, another drawback of the simplified model becomes obvious, since the diffusion occurs alternating in both directions within the experiment, the peak height is already lower than expected by the simplified model and is therefore not correctly reproduced.

The determined reaction rate of sodium sulfite is in the range of $k_{CR2} = 1 \cdot 10^4$ to $100 \cdot 10^4$ L·mol^{-1}·s^{-1} as described by Kück et al.. Within the literature, as already explained in state of

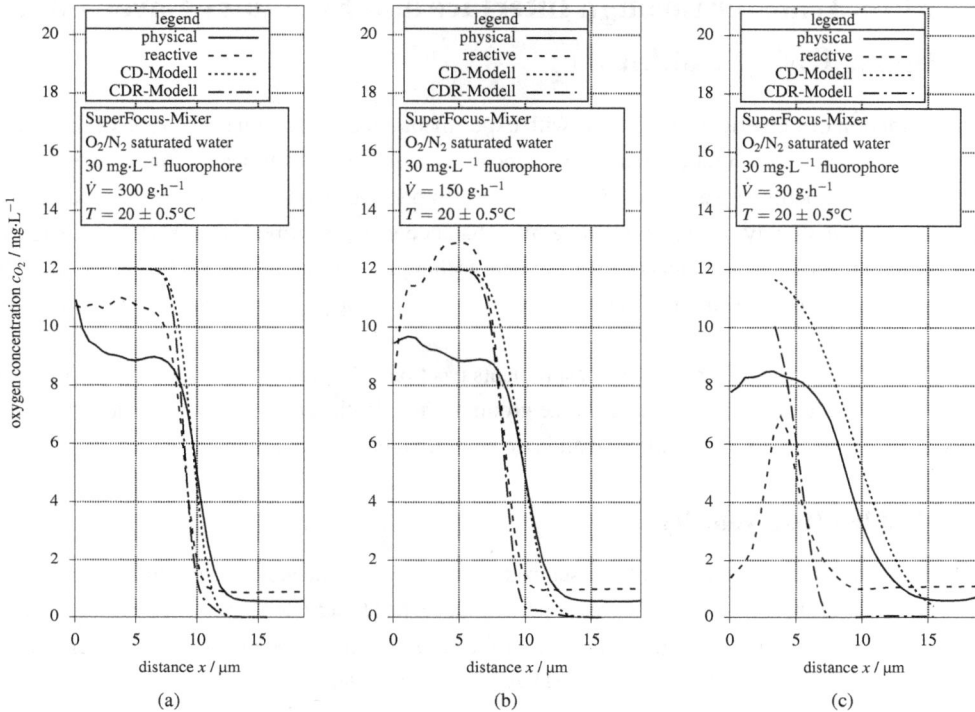

Figure 4.12 Comparison of the experimental physical and reactive mixing data with the CDR-model and convection/diffusion (CD) model at different flow rates. (a) flow rate of 300 g·h^{-1}, (b) 150 g·h^{-1}, (c) 30 g·h^{-1}.

the art, there are several reaction rates and orders available e.g. [Lan97, Erm02, Her03, Kar08, Shu12, She12, Liu15], so that the determined reaction rate is not finally confirmed.

It can be concluded, that the combination of CLSM measurements and numerical simulations is an excellent tool for the determination of intrinsic kinetics. With a further application of the simplified CDR model an approximation of the reaction kinetics is possible before a numerical simulation is performed, which will lead to a faster determination.

4.2 Mass Transfer through Interface and Boundary Layer - Rectilinear Rising Bubbles

For the validation of numerical simulations with experimental results and quantification of the transferred mass, accurate concentration field measurements are required. Rectilinear rising bubbles with a symmetric wake structure are chosen and investigated. In previous work (see section 2.2) rectilinear bubbles have been already investigated and a mass balance was performed [Küc09, Küc10, Küc11, Jim13]. Due to recent developments of measurement techniques, limitations like frame rate and laser sheet structure have been overcome to obtain the detailed development and structure of the bubble wake.

Within the following section, p-LIF measurements (section 3.2.2) are evaluated and the results are used to validate numerical simulations from the group of Prof. Bothe at the TU Darmstadt. Therefore besides mass transfer, also the bubble size and rise velocity are evaluated.

4.2.1 Bubble Rise Velocity

It is likely, that the fluorescent dye for the visualization of oxygen influences the mobility of the bubble interface and therefore also the mass transfer performance, as already described (section 2.2.1). For an accurate numerical simulation, the degree of surface mobility has to be quantified by determining the bubble size and rise velocity with a MATLAB® routine. Due to a low gray level gradient of bubble interface and surrounding liquid, an edge detection is impossible, so the bubble shadow is used to identify the bubble size and rise velocity. Therefore, first a threshold is defined manually and afterwards all recordings are evaluated. The bubble diameter is determined by assuming a spherical shape, which is valid for bubbles below 1 mm in diameter [Cli78]. Figure 4.13 shows the determined bubble diameter and velocity, as well as the velocity prediction described by Tomiyama [Tom98] for a clean, slightly and full contaminated system (see table 2.2) and additional literature data and correlations.

It can be clearly observed, that a major influence of the fluorophore on the bubble rise velocity exists. The velocity is slightly higher than the predicted rise velocity for a full contaminated system, but is lower than the rise velocity in case of a slightly contaminated system, so that the mass transfer is likely very similar to a rigid sphere. The contamination has to be taken into account within all numerical simulations of mass transfer processes evaluated with the p-LIF method. It is further assumed, that the fluorophore acts as a typical surfactant, which continuously is absorbing and desorbing on the interface to form a stagnant cap (see section 2.2.1).

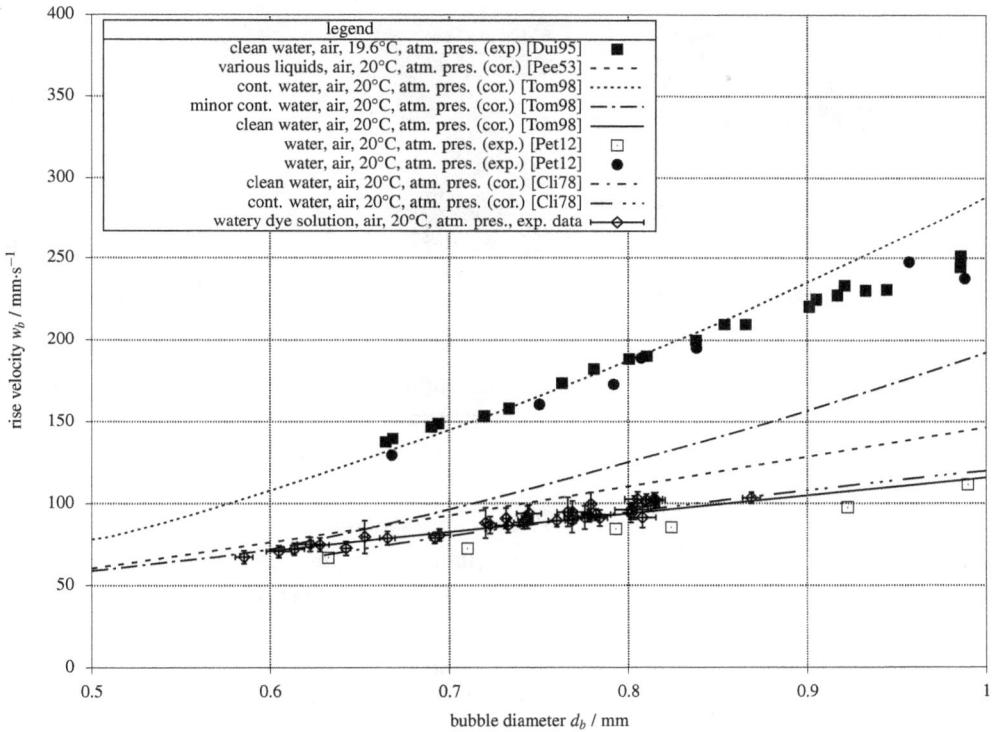

Figure 4.13 Rise velocity of rectilinear rising air bubbles within the fluorophore solution in comparison with different correlations and literature data.

4.2.2 Reactive Mass Transfer at Single Spherical Bubbles - Qualitative Data

Figure 4.14 a) - c) shows the averaged ROI images for rectilinear rising air bubbles with a diameter of d_b = 0.58, 0.68 and 0.78 mm. It is observed, that the wake structure directly behind the bubble is broader and elongated with rising bubble diameter, which has been already described in the literature [Was87] and is confirmed for the Reynolds number range of Re = 39 to 82. The rise velocities of the bubbles are w_b = 67, 79 and 93 mm·s^{-1} with a mean standard deviation of $\pm$ 5 mm·s^{-1}.

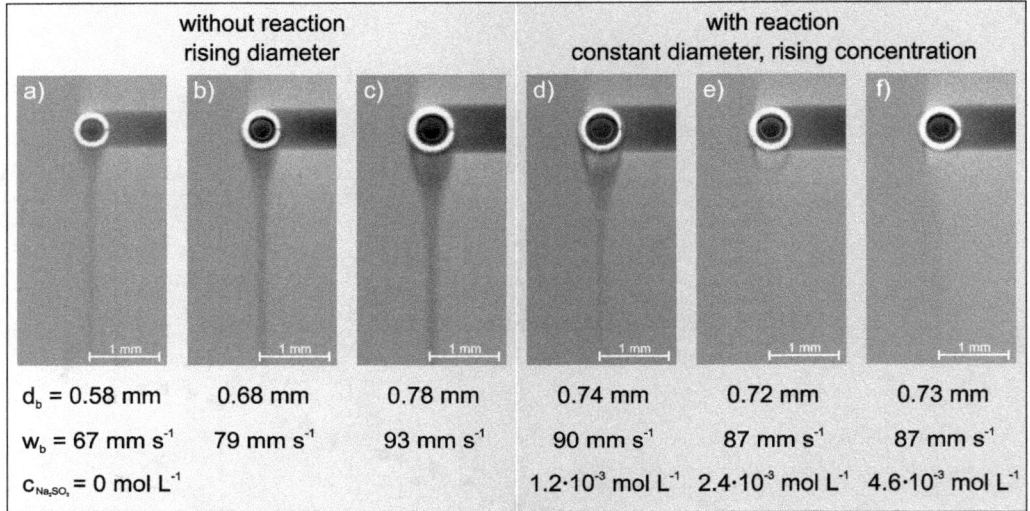

without reaction rising diameter			with reaction constant diameter, rising concentration		
a)	b)	c)	d)	e)	f)
d_b = 0.58 mm	0.68 mm	0.78 mm	0.74 mm	0.72 mm	0.73 mm
w_b = 67 mm s^{-1}	79 mm s^{-1}	93 mm s^{-1}	90 mm s^{-1}	87 mm s^{-1}	87 mm s^{-1}
$c_{Na_2SO_3}$ = 0 mol L^{-1}			1.2·10^{-3} mol L^{-1}	2.4·10^{-3} mol L^{-1}	4.6·10^{-3} mol L^{-1}

Figure 4.14 Bubble size and sulfite concentration influence on the oxygen concentration field.

Additionally, to physical mass transfer, also reactive mass transfer with a superimposed oxidation of sodium sulfite is investigated for a comparison with numerical investigations. Figure 4.14 d) to f) shows the obtained ROI images of rising air bubbles with a diameter d_b of $\approx$ 0.73 mm, a rise velocity of $w_b \approx$ 88 mm·s^{-1} and different sodium sulfite concentrations. The residual concentration of oxygen is varying slightly between the investigations and is in case of physical mixing approximately 0.1 mg·L^{-1}. In case of the superimposed sodium sulfite oxidation all oxygen is consumed by the reaction, so that it results in a residual concentration of 0 mg·L^{-1}. Due to these differences in oxygen concentration, the fluorescence intensity is varying slightly within the recordings additionally to the influence of temperature and pH value.

With rising sodium sulfite concentration, a drain of the wake structure is observed (see figure 4.14 d-f). At small sulfite concentrations the oxygen within the stagnation ring at the rear of the bubble is already drained and the wake itself is thinned by the reaction (figure 4.14 d). By increasing the reaction rate such as, for instance with increasing sulfite concentration (this work), or increasing temperature [Küc11], the wake structure is more and more drained. With doubling the sodium sulfite concentration, oxygen is observed only at the separation point and within the following streamlines (figure 4.14 e). With a further increase of the reaction rate, the transferred oxygen is consumed very close to the interface (figure 4.14 f) and is only observable at the separation point.

4.2.3 Reactive Mass Transfer at Single Spherical Bubbles - Quantitative Data

To determine the mass transfer coefficient, a quantitative evaluation of the transferred mass, as a mass flow rate $\dot{M}$, is required. The analysis is based on the work of Kück *et al.* [Küc09, Küc10, Küc11] and Jimenez *et al.* [Jim13]. At first an evaluation of the concentration field, as described in section 3.1.2, is performed. Figure 4.15 shows the oxygen concentration fields as pseudo color images. These images are already post processed for the transferred mass evaluation. Therefore, the bubble is masked and all values below 0.1 mg·L^{-1} are set to not a number (NaN), which is represented in the color images as a white color. Additionally, the images are cropped to a size of 80 x 738 pixel, so the bubble and the bubble wake are cut along their symmetry axis by additionally neglecting the bubble shadow side.

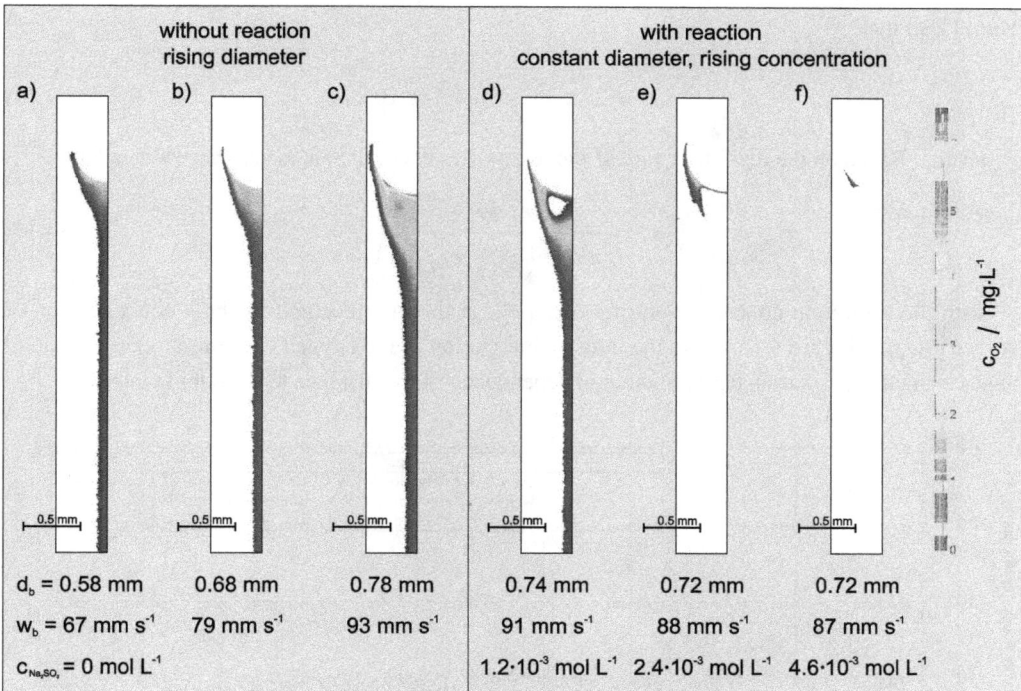

Figure 4.15 Influence of bubble size and sulfite concentration on the oxygen concentration field as pseudo color image, white as NaN.

The evaluation of the transferred mass is performed in case of physical mass transfer only, since the consumed oxygen amount can not be determined in case of the sodium sulfite reaction. The mass flow rate of oxygen is defined as the product of volume flow rate $\dot{V}$ and mass concentration c_{O_2} of oxygen

$$\dot{M} = \dot{V} \cdot c_{O_2}. \tag{4.11}$$

While the oxygen concentration is determined by the LIF measurements [Küc11], the volume flow rate is approximated. It is assumed, that the liquid velocity v_l within the wake is equal to the rise velocity of the bubble w_b [Jim13], so that

$$\dot{M} = w_b \cdot A \cdot c_{O_2} \qquad (4.12)$$

with A as the cross section perpendicular to the rise direction results [Jim13]. The transferred mass within this cross section can be determined from the two-dimensional recording by solving the surface integral in cylindrical coordinates and balancing the concentration within the considered plane $(c_{O_2}(r, \varphi))$

$$\dot{M} = w_b \int_r \int_\varphi c_{O_2}(r, \varphi) \cdot r \cdot d\varphi dr. \qquad (4.13)$$

Due to the rectilinear bubble rise, a high rotational symmetry exist and the two-dimensional plane of the recordings lead to a simplification of equation 4.13 by the assumption, that $c_{O_2}(\varphi)$ is constant [Küc11], so that

$$\dot{M} = 2 \cdot \pi \cdot w_b \int_r c_{O_2}(r) \cdot r \cdot dr \qquad (4.14)$$

results. Based on the mass flow rate $\dot{M}$ the mass transfer coefficient k_L can be evaluated as

$$k_L = \frac{\dot{M}}{\pi d_b{}^2 \cdot \left(c_{O_2}^* - c_{O_2,bulk} \right)} \qquad (4.15)$$

with the saturation concentration of oxygen $c_{O_2}^*$ at the interface and the bulk concentration of oxygen $c_{O_2,bulk}$. Figure 4.16 shows the through integration of the oxygen concentration in the lateral planes evaluated mass transfer coefficient in dependency of the distance to the bubble interface.

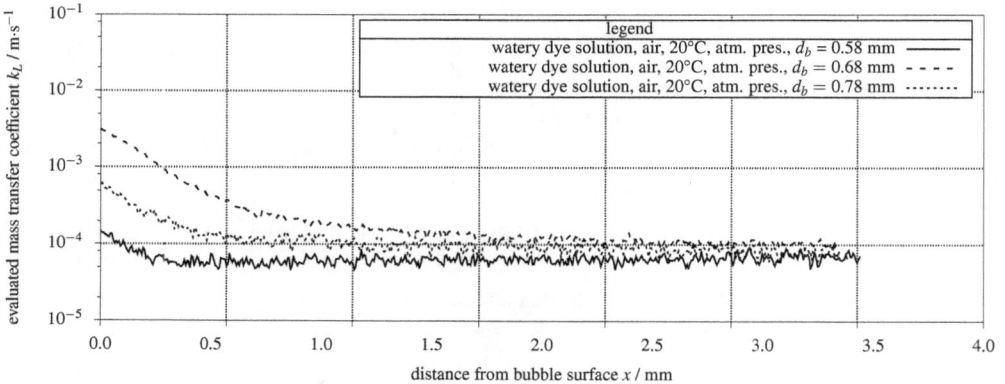

Figure 4.16 Through integration of the oxygen concentration in the lateral planes evaluated mass transfer coefficient for the bubble diameter $d_b = 0.58$, 0.68 and 0.78 mm in axial direction to the bubble interface.

As already observed by Kück et al. [Küc10], the mass transfer evaluation within the stagnation vortex behind the rising bubbles fails due to neglecting the velocity field. Therefore, the mass transfer coefficient evaluation is performed at a distance above one bubble diameter. For the determination within this work, the last 250 lateral planes are averaged. Figure 4.17 shows the experimental mass transfer coefficient in dependency of the bubble diameter. A scattering between the different experimental investigations exists, while the mean mass transfer coefficient of $1.06 \cdot 10^{-4}$ m·s^{-1} is in the same magnitude as the mass transfer coefficient of $2.02 \cdot 10^{-4}$ m·s^{-1} for a 0.91 mm bubble found by Kück *et al.* [Küc10]. The investigations of Kück *et al.* are performed in nearly the same material system and with a similar p-LIF technique. There are two differences within the experimental investigation. First of all, only one snapshot of the bubble is evaluated, so that through background deviations and noise a high deviation within the determined mass transfer coefficients exist. Furthermore, the viscosity of the fluorophore solution was increased to enable a better comparability to the numerical simulations, which is today no longer necessary and therefore the bubble diameter is smaller.

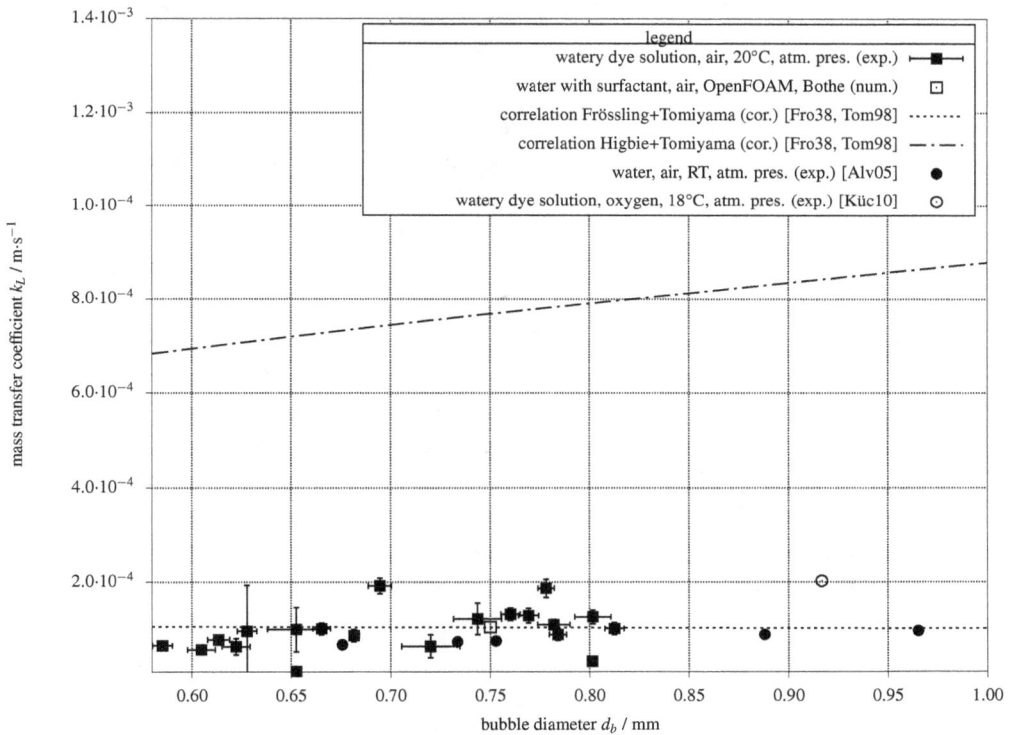

Figure 4.17 Comparison of the experimental and numerical determined mass transfer coefficients with correlations from literature.

The found mass transfer coefficients agree well with the correlation of Frössling [Fro38] for rigid particles in combination with the rise velocity of contaminated bubbles according to Tomiyama [Tom98], which proves again the surfactant behavior of the fluorophore. Alves *et al.* [Alv05] investigated the mass transfer for bubbles in the diameter range of 0.5 to 5 mm with the bubble shrinkage method in distilled water with and without surfactant. The values for a fully immobilized surface or respectively for rigid particles measured by Alves agree well with the experimental results within the fluorophore solution. While the numerical simulations of the group of Prof. Bothe at

TU Darmstadt shows an excellent agreement with the experimental results and the correlations of Frössling [Fro38].

To conclude, the determination of mass transfer coefficients with high-speed p-LIF in case of rectilinear bubbles is an excellent tool for the determination of mass transfer coefficients. Based on the averaging procedure of several ROI, the signal-to-noise ratio is significantly improved, so that a more accurate determination in comparison with [Küc09, Küc10, Küc11, Jim13] is possible. Additionally, this technique allows a precise determination of the bubble diameter and rise velocity and revealed also, that the fluorophore for the visualization of the oxygen concentration field has a major influence on the mobility of the bubble surface, and therefore on rise velocity and mass transfer performance.

4.3 Influence of Boundary Layer Deformation - Bouncing Bubbles

The hydrodynamic in bubbly flows is mainly influenced by bubble-induced turbulence and bubble-bubble interactions. As a result also the gas-liquid mass transfer and mixing of reactants is influenced. In case of a parallel/consecutive reaction, where the timescale of the chemical reaction is in the same order than mixing and mass transfer, the yield and selectivity will be affected by local hydrodynamics. Therefore, specific designed hydrodynamic conditions will lead to optimization of yield and selectivity.

Nevertheless, the interdependence between mass transfer, mixing and chemical reaction in bubbly flows is so far not understood and a reliable prediction is impossible. To gain a deeper knowledge on the interdependence of these timescales, within this section the influence of different parameters on mass transfer and mixing in case of bubble bouncing, within wake structures and at wobbling bubbles is investigated. To identify the possible influence of bubble bouncing on mass transfer, high-speed p-LIF investigations of the concentration field around bouncing oxygen bubbles with and without superimposed sulfite oxidation are performed to get a deeper insight into the additional mixing effect. As described in section 3.3 the direct interaction of a mechanical fixed bubble with a free rising bubble is induced. Additionally, to the p-LIF recordings, the shrinkage of CO_2 bubbles in deionized water is investigated to quantify the mass transfer and enhancement.

4.3.1 Influence on Local Mass Transfer and Reactions

To reveal the processes that enhance mass transfer in case of bubble bouncing, local concentration fields for air bubbles with p-LIF and with superimposed sodium sulfite oxidation are recorded (section 3.3.2). Here also a post-processing image is required. The procedure for raw data does not differ from the one described within section 4.2, except that no ROI is defined and no mean images are calculated. Figure 4.18 shows the qualitative concentration fields for four helical rising air bubbles with an equivalent volume diameter of $d_{eq,V} \approx 2.0$ mm and a rise velocity of $w_b \approx 32$ cm$\cdot$s^{-1}.

The saturated concentration wake of a helical rising air bubble in oxygen desorbed solution is shown in figure 4.18 a). Figure 4.18 b), c) and d) shows the concentration wake in case of the superimposed reaction. In case b) and c) a sodium sulfite concentration of $4.6 \cdot 10^{-3}$ and $5.95 \cdot 10^{-3}$ mol$\cdot$L^{-1} is superimposed. Since the concentrations are very similar to the one used in section 3.1, the reaction rate is in the same magnitude. As expected, the observation shows the reaction consumes

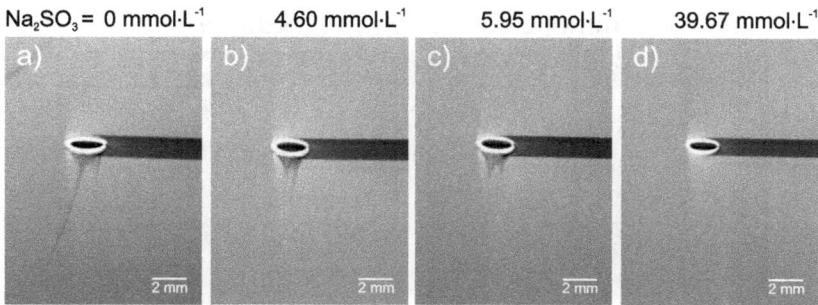

Figure 4.18 Qualitative local concentration fields for helical rising bubbles with and without superimposed sodium sulfite oxidation [Tim16].

oxygen slightly already at the boundary layer but mainly in the bulk phase. At very high sodium sulfite concentrations (figure 4.18 d), no oxygen wake is observed. The reaction rate here is higher than the mass transfer rate, which leads to a consumption of the transferred oxygen in the vicinity of the boundary layer and therefore no oxygen is visible within the recording.

From these preliminary investigations it is chosen to use the concentration of $5.95 \cdot 10^{-3}$ mol·L^{-1}, two times the concentration investigated in section 3.1, for the following investigations at bouncing bubbles since a clear influence of the reaction rate is observable. To precipitate reproducible bouncing the upper capillary is now placed in the trajectory of the rising bubble. Figure 4.19 shows snapshots from recordings of physical (a,b) and reactive (c,d) mass transfer 30 and 60 ms after bouncing. In both cases a detached concentration wake that forms an oxygen eddy is observed (a and c). While the oxygen eddy in case of physical mass transfer is propagating through the volume and is dissipating over time (b), the oxygen eddy in the reactive case (compare c and d) is consumed, so that the decay of the eddy is not observed. The calculated equivalent volume diameter $d_{eq,V}$ for the fixed bubble within these investigations is 2.42 ± 0.06 mm and 2.03 ± 0.05 mm for the rising bubble with a velocity in average of 31.8 cm·s^{-1}.

The formation of a concentration eddy from the detached wake, indicates a high mixing intensity close to the boundary layer induced by the bouncing process. It is observed, that the transferred species is consumed within the wake structure and therefore also after the detachment in case of bouncing for reactions in the range of $k_{CR2} = 1 \cdot 10^4$ to $100 \cdot 10^4$ L·mol^{-1}·s^{-1} [Küc11]. Additionally, to the shown wake detachment in case of bouncing, also a boundary layer deformation of both bubbles occurs, so that a surface stretching and detachment of smaller eddies is observed. Both will also contribute to mass transfer enhancement and mixing as already described by Gläser [Gla77]. For the desired case of a parallel/consecutive reaction, where the product distribution is determined by the degree of mixing [Roe01], an influenced yield and selectivity will result [Tim16].

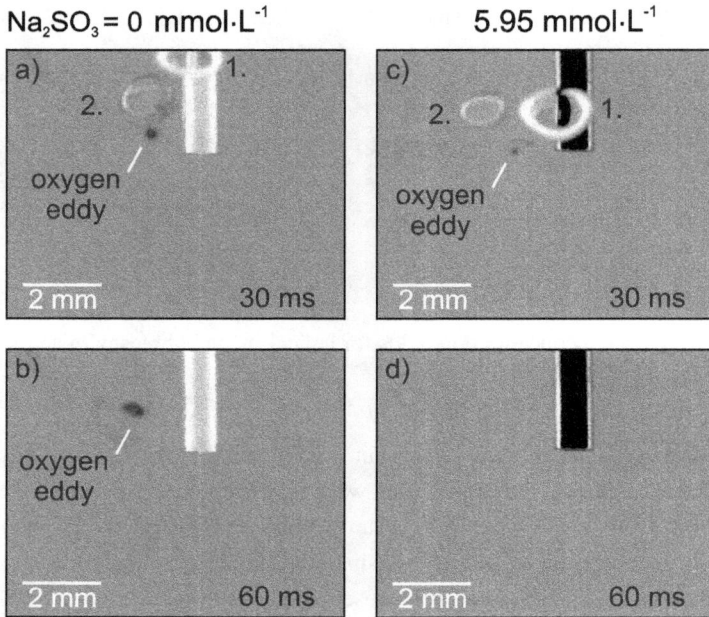

Figure 4.19 Snapshots of the recorded oxygen concentration fields for a mechanically fixed bubble (1.) and rising bubble (2.) with physical (a,b) and reactive (c,d) mass transfer 30 and 60 ms after bubble bouncing [Tim16].

4.3.2 Influence on Overall Mass Transfer

Figure 4.20 shows snapshots of the recording from a shrinking CO_2 bubble in DI water. To determine the transferred mass, a procedure according to Hosoda et al. [Hos14] and Kastens et al. [Kas15] in MATLAB® is developed.

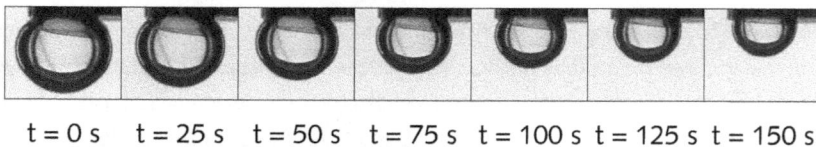

$$t = 0\,\text{s} \quad t = 25\,\text{s} \quad t = 50\,\text{s} \quad t = 75\,\text{s} \quad t = 100\,\text{s} \quad t = 125\,\text{s} \quad t = 150\,\text{s}$$

Figure 4.20 CO_2 bubble shrinkage in DI water with no collision. Pictures from [Rud15]

By using edge detection, the recorded image is transformed into a binary image (figure 4.21) and assuming rotational symmetry, the bubble volume is calculated by a segmentation of the image. Each segment has the height of one pixel (Δz) and the volume of a cylinder

$$V_B = \sum_{i=1}^{n} \pi r_i^2 \Delta z \tag{4.16}$$

with V_B as bubble volume and r_i as segment radius.

The bubble is mechanically fixed with a stainless steel rod and is therefore deformed. From the recordings it is assumed, that an ellipsoidal cap in good agreement determines the bubble surface area.

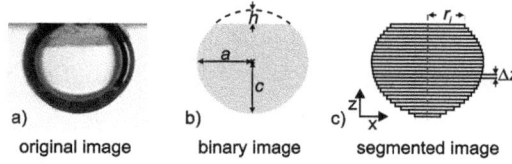

a) original image   b) binary image   c) segmented image

Figure 4.21 Image processing procedure for CO_2 bubble shrinkage [Tim16].

Therefore, also the semi-axes a, c and the height of the missing cap h are calculated by MATLAB® for the identified bubble shape. The surface area of the cap A_{surf} is approximated by the surface area of a full ellipsoid $A_{ellip.}$ and subtracted by the area of the missing cap A_{cap}

$$A_{surf} = A_{ellip.} - A_{cap} = 4\pi \left(\frac{(a^2)^{\frac{8}{5}} + (2ac)^{\frac{8}{5}}}{3} \right)^{\frac{5}{8}} - 2\pi ah . \tag{4.17}$$

The area in contact with the fixation device is neglected, since this area does not contribute to mass transfer. The evaluated surface equivalent diameter is smaller than the real bubble diameter and is calculated according to

$$d_{eq,A} = \sqrt{\frac{A_{surf}}{\pi}} \tag{4.18}$$

The code is verified by using images of a stainless steel sphere and an error of less than 1% for diameter, surface area and volume is achieved [Tim16].

Overall Mass Transfer Coefficient Evaluation

The overall mass transfer coefficient k_L is evaluated from the decrease of bubble equivalent surface diameter with a few assumptions.

The change in amount of substance for the CO_2 bubble is described by

$$\frac{dn_m}{dt} = -k_L \cdot A_{surf} (c^* - c_{bulk}) \tag{4.19}$$

with c^*, as the CO_2 concentration at the bubble interface and c_{bulk}, as the CO_2 concentration within the bulk phase. The CO_2 concentration at the interface is defined by Henry's law as

$$p = \frac{c^*}{c_{bulk} + c^* + c_v} \cdot H \tag{4.20}$$

where p is the pressure inside the bubble, c_v the molar concentration of water (55.5 kmol·m^{-3}) and H the Henry constant. For the evaluation it is assumed, that the CO_2 concentration within the bulk is zero due to the saturation with nitrogen and desorption of carbon dioxide within the bubble column. By solving equation 4.20 with the assumption of atmospheric conditions, a value of 39.6 mol·m^{-3} for c^* is obtained. Therefore, c_v/c^* is approximately 1400 and it follows, that $c^* + c_v$ is nearly equal to c_v, so that 4.20 is simplified by neglecting the bulk concentration to

$$c^* = \frac{c_v \cdot p}{H}. \tag{4.21}$$

The pressure inside the bubble is defined as

$$p = p_a + \rho_L g h_{(z)} \tag{4.22}$$

p_a is the atmospheric pressure, ρ_L the liquid density, g the gravitational acceleration and $h_{(z)}$ the height of the supernatant fluid. Since the bubble radius is above 1 mm, the Young-Laplace pressure is neglected in equation 4.21. With inserting equation 4.21 and the converted equation 4.18 into equation 4.19 yields

$$\frac{dn_m}{dt} = -k_L \pi d_{eq,A}^2 \frac{c_v p}{H}. \tag{4.23}$$

The decrease in molar mass of CO_2 is described by assuming CO_2 as an ideal gas and neglecting counter diffusion of CO_2 into the bubble

$$\frac{dn_m}{dt} = \frac{\pi}{6RT} \frac{d\left(p d_{eq,A}^3\right)}{dt} \tag{4.24}$$

where R is the universal gas constant and T the liquid temperature.

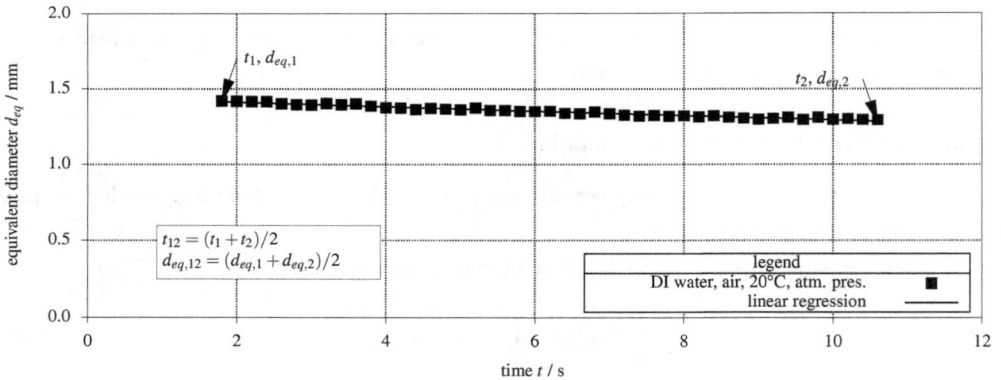

Figure 4.22 Example of linear regression for a shrinking CO_2 bubble in DI water [Tim16].

By evaluating $d p d_{eq,A}^3 / dt$ with a centered difference scheme, equating the right-hand sides of equation 4.23 and 4.24 and conversion to k_L results

$$k_L = \frac{H \left(p_2 d_{eq,A,2}^3 - p_1 d_{eq,A,1}^3 \right)}{6RT c_v \overline{p_{12}} \left(\overline{d_{eq,A,12}} \right)^2 (t_1 - t_2)} \tag{4.25}$$

with 1 and 2 as indices for the time t_1 and t_2. $\overline{d_{eq,A,12}}$ and $\overline{p_{12}}$ are the arithmetic averages for pressure and diameter at time t_1 and t_2 [Kas15]. From the recording, the bubble diameters during the shrinkage are evaluated and with a linear regression (see figure 4.22), the parameters for t_1 and t_2 are

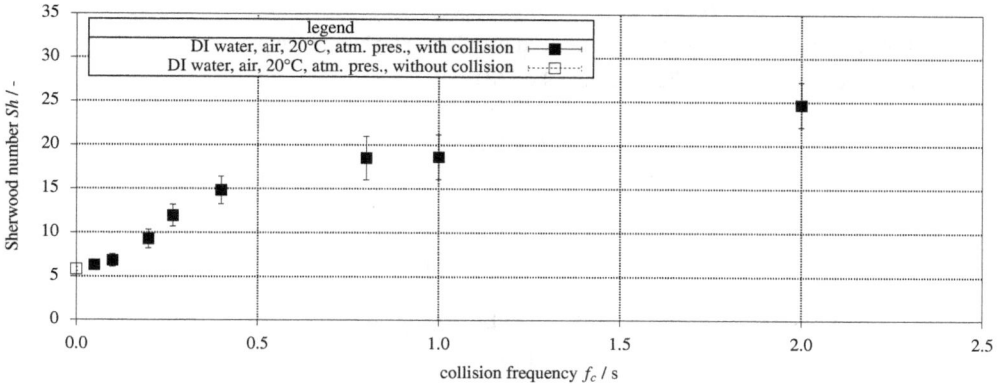

Figure 4.23 Influence of bouncing on the overall mass transfer [Tim16].

determined. Due to the mechanical fixation, the bubble position relative to the liquid surface and as a result the pressure is constant during the experiment, equation 4.25 simplifies to

$$k_L = \frac{H \left(d_{eq,A,2}^3 - d_{eq,A,1}^3 \right)}{6RT c_v \left(\overline{d_{eq,A,12}} \right)^2 (t_1 - t_2)} \tag{4.26}$$

By applying the calculated mass transfer coefficient into equation 2.23 and using the diffusion coefficient D_L for CO_2 in water, the Sherwood number Sh_e is calculated. Figure 4.23 shows the measured Sherwood numbers Sh at different collision frequencies f_c. A strong dependency of the collision frequency between the fixed and the rising bubble is found.

The theoretical Sherwood number of $Sh = 2$ is not reached within these investigations for a bubble without collision, which indicates, that the liquid is not stagnant as assumed. Most likely an induced convection resulting from a temperature or concentration gradient or from previous bubble movement occurs. Therefore, a slight convection is affecting the overall mass transfer in all cases and can not be separated from the influence of bouncing to the overall mass transfer. However, a significant mass transfer enhancement is found and has to be considered for bubbly flows.

4.3.3 Modeling of Binary Bubble Interactions

Based on the observations at bouncing bubbles, different phenomena are identified, which influence mass transfer in case of binary bubble interactions and are used to develop a model concept for the rising bubble. While the bubble is rising uninfluenced and freely, a concentration wake is developed. The concentration within this wake is determined by mass transfer and residence time at the boundary layer by the velocity field. In case of an additional chemical reaction, also the reaction rate determines the concentration field. Mixing of this saturated wake occurs mainly through diffusion (figure 4.24 a). When bouncing occurs, the concentration wake is detached and forms eddies within the bulk phase (b). These eddies lead to a region with a higher degree of mixing. In case of a parallel/consecutive reaction this area of higher mixing will influence the yield and selectivity. Additionally, to the detaching, also the boundary layer is deformed. This deformation additionally induces eddies through an acceleration and deceleration of the fluid phase, which also enhance mass transfer (c). The stable form of the observed bubble is an ellipsoid, so that the induced oscillation is damped and a settling of the boundary layer takes place and no more eddies are detached (d).

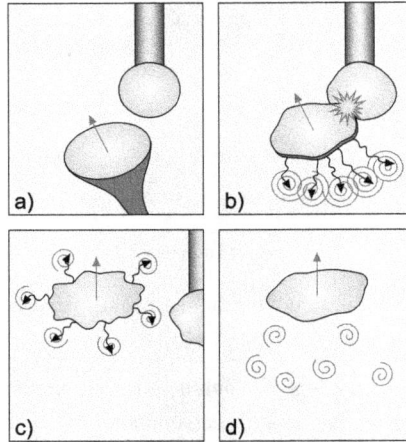

Figure 4.24 Model concept to describe the influence of bouncing on mass transfer [Tim16].

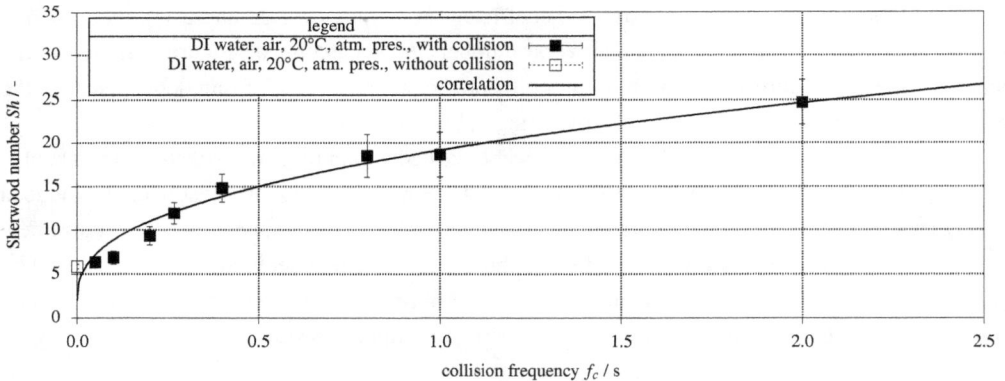

Figure 4.25 Influence of bouncing on the overall mass transfer [Tim16].

To model the mass transfer performance, the empirical equation 2.25 according to Anderson [And67] is modified to fulfill the solution of $Sh = 2$ for a spherical gas bubble at rest [Kra12] and using a least square approach by changing the pre-factor and power of equation 2.25. This leads to equation 4.27

$$Sh_e = 2 + 0.91 \left(\frac{d_{eq,A}^2 f_c}{D_L} \right)^{2/5} \tag{4.27}$$

The identified change in pre-factor and power probably results from the mechanical fixation of the bubble, so that a higher damping of the oscillation occurs and therefore wake shedding is hindered.

While the hydrodynamical processes of bouncing are already described within the literature by several authors e.g. [Dui94, Dui95, San09], the influence of bouncing on mass transfer is not described yet. Sanada *et al.* visualized already the bubble wake in case of collision through a dyed wake [San09]. It is observed for side by side bouncing, that the bubble wake is detached during the collision (see also figure 2.6). The investigations performed within this work illustrate the influence of bubble bouncing on mass transfer in a more detailed manner and includes the influence of a chemical reaction.

An enhancement of mass transfer performance in case of bubble interactions is proven within this work and a semi-empirical equation is developed, which describes the influence of the collision frequency (published in [Tim16]). The developed equation is based on the work of Anderson [And67] for the description of mass transfer of shape oscillating bubbles. It is observed, that the collisions also lead to shape oscillations at the fixed bubble, so that a similar approach is chosen.

P-LIF investigations are performed to clarify the local processes that contribute to mass transfer enhancement. This investigations showing a detachement of the bubble wake if a bubble collision occurs and small eddies are formed. Hlawitschka *et al.* investigated reactive bubble bouncing with a very similar experimental setup and observed also, that the bubble wake is detaching simultaneously with the bubble collision and is afterwards forming a new bubble wake, while the former wake is dissolving within the solution [Hla17]. These investigations were performed with background illumination high speed recordings, so that no concentration field data are available, while the mass transfer enhancement was also not investigated.

The performed investigations close a gap in the knowledge of the mass transfer processes involved in the bouncing procedure and allow a first approximation of the influence of bouncing in bubbly flows with the semi-empirical equation.

4.4 Influence of Mixing within the Bubble Wake - Taylor Bubble

To study the influence of the gas-liquid mass transfer on chemical reactions, Taylor bubbles are used within this work. By the use of background illumination investigations during bubble shrinkage the mass transfer is quantified and different wake structures are identified. Additionally, first investigations with a p-LIF approach on a reactive system with self indication are conducted. Nevertheless, the defined conditions in case of Taylor bubble do not occur in a fully developed bubbly flow. The characterization of wake structures in bubbly flows is not trivial, since the structure is three-dimensional and time dependent. So far only few techniques provide detailed information in such a case. Within this work TRS-LIF (section 3.5.1) is further developed and used to identify characteristic structures for physical mass transfer.

4.4.1 Mass Transfer Enhancement at Reactive Taylor Bubbles

Through the constant rise velocity and the possibility of investigations over a time range of several minutes, Taylor bubbles allow a relatively easy determination of the mass transfer coefficient as shown by Kastens *et al.* [Kas15]. This technique allows also in case of reactive mass transfer the determination of the reactive mass transfer coefficient and the enhancement factor E (see section 2.2.4).

In a first approach, the oxidation of sodium sulfite, as already used before, is applied to evaluate the enhancement factor and the mass transfer coefficient. The determination is similar to the described evaluation in section 4.3.2. Additionally, the bubble diameter is evaluated for a spherical bubble with the same surface area as the Taylor bubble. Figure 4.26 shows images from the dissolution process of an oxygen bubble in water.

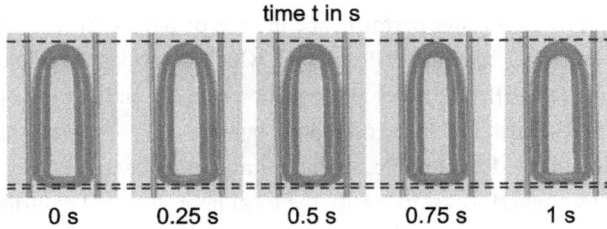

Figure 4.26 Images of the dissolution of oxygen in water over time, adapted from [Com17].

The dissolution process, barley visible within the images, is slower as for CO_2 bubbles (compare section 4.3.2), as one could expect. Therefore only a small decrease of the equivalent diameter $d_{b,eq}$ over time is recognized. Nevertheless, this small decrease enables in combination with equation 4.26 to calculate the mass transfer coefficient.

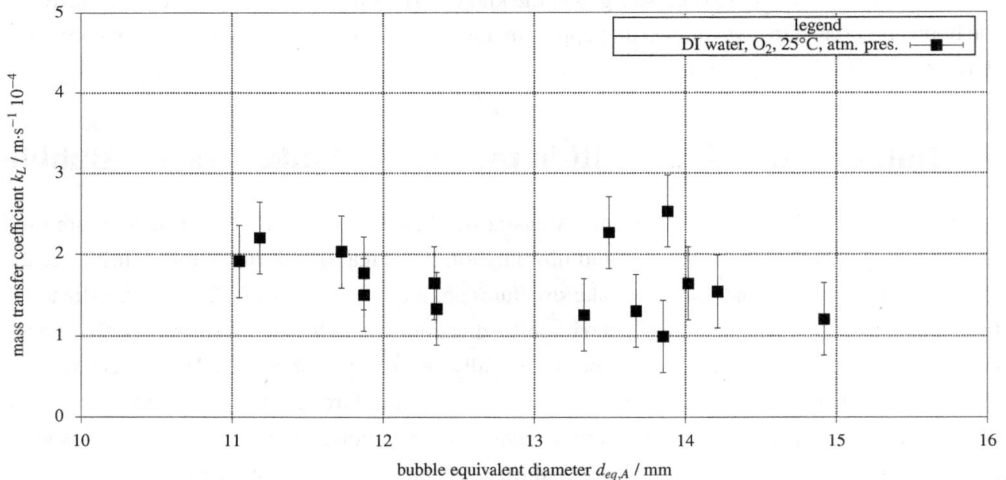

Figure 4.27 Mass transfer coefficient for the absorption of O_2 in water (8 mm capillary). Data from [Com17].

Figure 4.27 illustrates the determined mass transfer coefficients as a function of the bubble equivalent diameter. It is observed, that the mass transfer coefficient is, as well as the rise velocity, independent from the bubble equivalent diameter and in average $k_L^0 = 1.71 \cdot 10^{-4} \pm 9.68 \cdot 10^{-5}$ m·s^{-1}. The reactive mass transfer coefficient k_L^R and the enhancement factor are determined with a solution of sodium sulfite ($c_{Na_2SO_3} = 0.04$ mol·L^{-1}). As expected, the bubble dissolution is much faster through the reaction. The difference in shrinkage velocity is shown in figure 4.28.

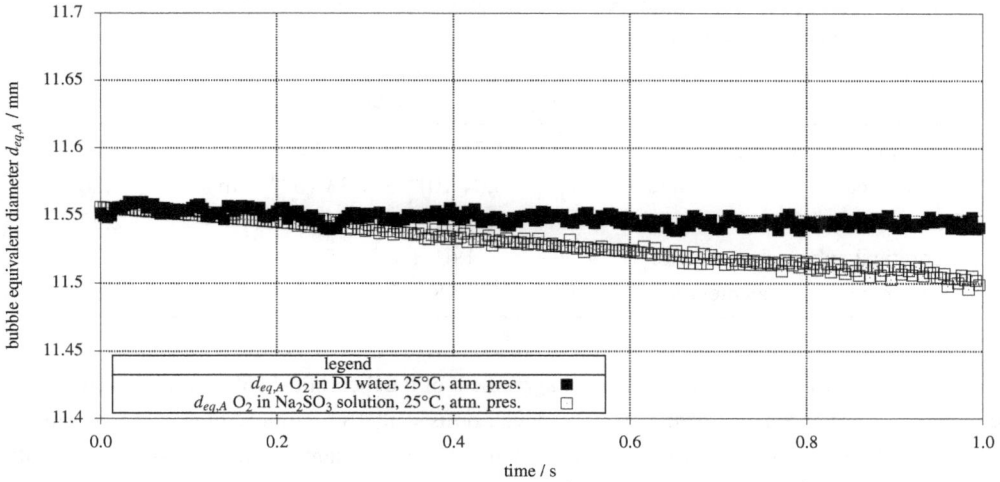

Figure 4.28 Dissolution of oxygen in water and sodium sulfite solution over time. Data from [Com17].

Here also different bubble sizes are investigated and evaluated according to equation 4.26 to obtain the reactive mass transfer coefficient. The results are compared with the physical mass transfer coefficients in figure 4.29. The average reactive mass transfer coefficient is $k_L^R = 3{,}04 \cdot 10^{-4} \pm 5.52 \cdot 10^{-5}$ m·s^{-1}. Table 4.2 summarizes the applied equations and results of the dissolution of oxygen in water and in sodium sulfite solution.

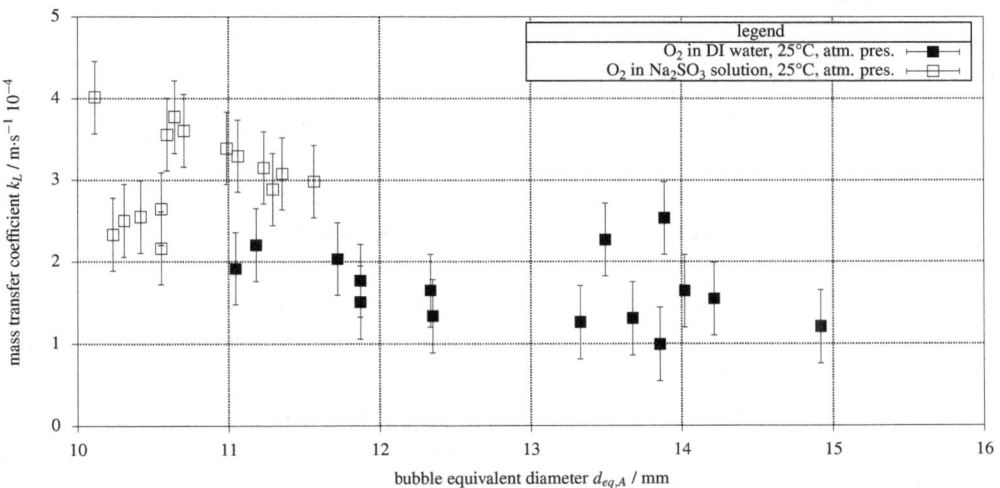

Figure 4.29 Reactive mass transfer coefficient for the absorption of O_2 in sodium sulfite solution (8 mm capillary). Data from [Com17].

As expected, the reactive mass transfer coefficient is higher than the physical mass transfer coefficient, while the deviation is higher as well. The determined Hatta number is in the range of $0.2 < Ha < 3$ for an intermediate reaction rate, as expected by the determined reaction rate

Table 4.2 Listing of k_L^0, k_L^R, Ha, E, and E^* for the reaction of O_2 with sodium sulfite in a Taylor flow capillary (8 mm diameter). Data from [Com17]

	result	value	unit	equation
mass transfer coefficient	$\overline{k_L^0}$	$1,71 \cdot 10^{-4} \pm 9.86 \cdot 10^{-5}$	$m \cdot s^{-1}$	4.26
reactive mass transfer coefficient	$\overline{k_L^R}$	$3,04 \cdot 10^{-4} \pm 1.33 \cdot 10^{-4}$	$m \cdot s^{-1}$	4.26
Hatta number	$\overline{Ha}$	$1.41 \pm 8.29 \cdot 10^{-2}$	-	2.32
theoretical enhancement factor	$\overline{E}$	$1.61 \pm 6.20 \cdot 10^{-2}$	-	2.33
experimental enhancement factor	$\overline{E^*}$	$1.98 \pm 1.36 \cdot 10^{-1}$	-	2.28

(section 4.1.3) [Mie17]. Additionally, a good agreement of the theoretical enhancement factor E and the experimental factor E^* is observed. The deviation is with 18% slightly higher as expected by literature (max. 8% [Ast67]). It can be concluded, that the experimental setup of hydro dynamically fixed Taylor bubbles is an excellent tool for long term investigation on a reactive two phase flow.

4.4.2 Wake Structures of Reactive Taylor Bubbles in Watery Systems

The understanding of the interdependence of local hydrodynamics, mass transfer and chemical reaction requires a reproducible and very adjustable experimental investigation. Taylor bubbles, with the constant rise velocity enable a quasi steady state investigation with reproducible and well-defined wake structures. To characterize the different wake structures, experimental investigations with a solution of an aqueous copper(I)ammonia solution $[Cu(NH_3)_4]^+$ aq. in combination with oxygen as gaseous phase is used [Rol11].

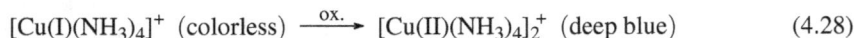

$$[Cu(I)(NH_3)_4]^+ \text{ (colorless)} \xrightarrow{\text{ox.}} [Cu(II)(NH_3)_4]_2^+ \text{ (deep blue)} \qquad (4.28)$$

If the colorless complex is oxidized, e.g. through atmospheric oxygen, the color is changing nearly instantaneously to deep blue, so that the wake structure is easily observed. Figure 4.30 shows for each capillary diameter the recorded color image (center) and compares the results with the combined PIV/LIF data of Kastens (left) [Kas17]. For a better visualization, the color pictures are post processed in that manner, that only the red channel from the RGB image is shown (right).

It is observed, that the wake structures are very similar for the case of dissolution of CO_2 [Kas17] (left) and for the reactive case of oxygen and $[Cu(NH_3)_4]^+$. For a channel diameter $d_p = 6$ mm and $Eo = 4.9$ (case a), a laminar wake structure is observable through the blue reaction product $[Cu(II)(NH_3)_4]_2^+$. The resulting structure is very similar to the PIV/LIF image of a CO_2 water system. For $d_p = 7$ mm (case b), the Eötvös number becomes higher $Eo = 6.7$ and leads to a toroidal vortex with a high concentration of product within the bubble wake. Here as well a good agreement with the PIV/LIF-images for the CO_2 water system results. In case (c), $Eo = 8.7$ and $d_p = 8$ mm, a turbulent wake is observed, that leads to a broad distribution of the of product nearly over the full cross section of the channel, which is again in good agreement with the PIV/LIF image.

It is shown, that different wake structures and therefore also the degree of mixing is adjusted by changing the channel diameter D and the Eötvös number Eo. The setup allows therefore investigations to adjust the yield and selectivity of parallel/consecutive reactions. Furthermore, it can be assumed,

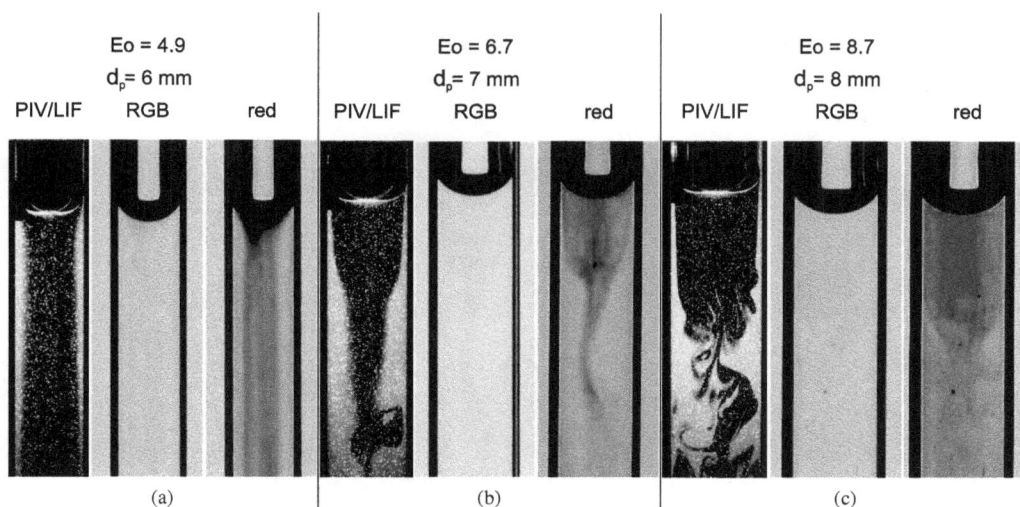

Figure 4.30 Comparison of wake structures at rising Taylor bubbles in channels in stagnant liquid. (a) d_p = 6 mm: laminar wake, (b) d_p = 7 mm: toroidal ring vortex, (c) d_p = 8 mm: turbulent wake with combined PIV/LIF data of Kastens [Kas17].

that even with changing the material system, e.g. solvents (viscosity, interfacial tension), the wake structure is reproducible through adjusting the Eötvös number by changing the pipe diameter [Kas17].

4.4.3 Wake Structures of Reactive Taylor Bubbles in Organic Systems

To overcome the influence of surfactants on the mass transfer performance in watery systems and to enable an detailed investigation of the reaction products, an organic reaction system from the SPP 1740 is required.

There are three reaction systems available within the SPP 1740 with different advantages and disadvantages, that are considered to identify an appropriate system. Figure 4.31 shows the three different reaction systems (Fe−NO, Fe−O_2 and Cu−O_2) and their properties. In case of these reaction systems, the small volume of the Taylor bubble experiment is a benefit, since only a small amount of reactant is available and a small substance usage for experimental investigations is essential.

All three systems show a color change with reaction progression for the qualitative evaluation and are observable with UV-Vis spectroscopy for the quantification of yield and selectivity. The solvent in case of Fe−NO is water and additionally the reactant is available from common suppliers so that cheap and relatively big scale experiments are possible. So far a monitoring of the reaction progression with p-LIF is not possible, since the coloration is very strong and no fluorescence answer is observed from this system. Due to the watery system, a surfactant influence has to be assumed. In the case of Fe−O_2 and Cu−O_2, organic solvents like methanol, acetonitrile or tetrahydrofuran are required. Nevertheless, this leads to reactants, which are designed on purpose and are not available from any supplier. So far the most promising reaction system is Cu−O_2 since it already allows a fluorescence investigation and the reaction rate is within the desired range.

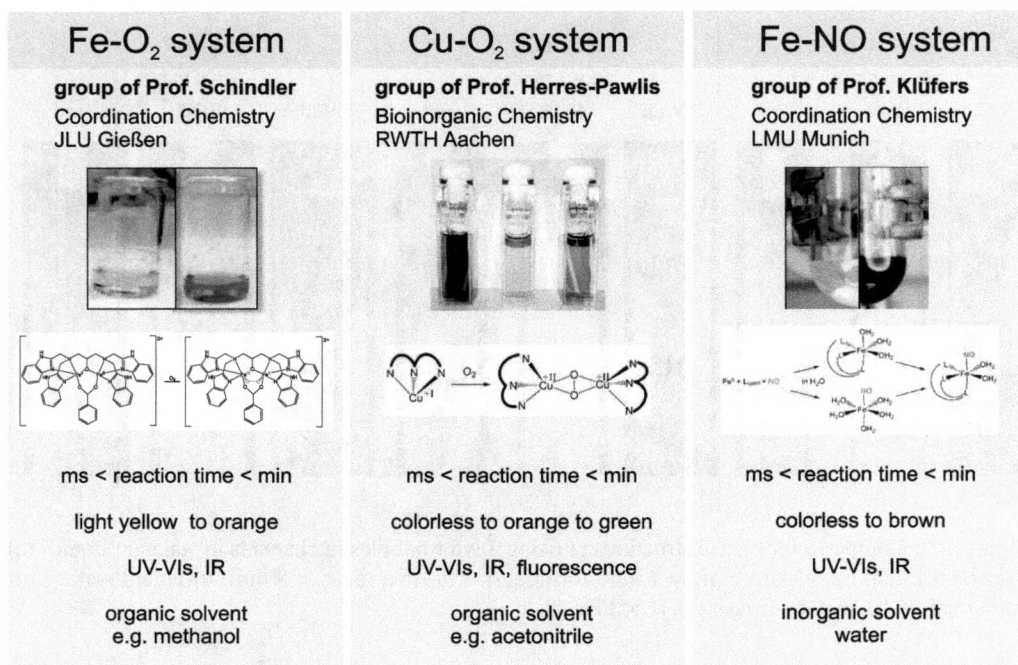

Fe-O₂ system	Cu-O₂ system	Fe-NO system

Fe-O$_2$ system

group of Prof. Schindler

Coordination Chemistry
JLU Gießen

Cu-O$_2$ system

group of Prof. Herres-Pawlis

Bioinorganic Chemistry
RWTH Aachen

Fe-NO system

group of Prof. Klüfers

Coordination Chemistry
LMU Munich

ms < reaction time < min	ms < reaction time < min	ms < reaction time < min
light yellow to orange	colorless to orange to green	colorless to brown
UV-VIs, IR	UV-VIs, IR, fluorescence	UV-VIs, IR
organic solvent e.g. methanol	organic solvent e.g. acetonitrile	inorganic solvent water

Figure 4.31 Reaction systems within the SPP 1740 and their properties.

This reaction system shows two reaction steps with different colorization. In a first step a orange bis(μ-oxo) complex is formed and in a second step it is decomposing to a green bis(μ-hydroxo) complex [Sch16]. The reaction rate for the first step is $10 \ \text{s}^{-1}$ and for the second step $1.5 \ \text{s}^{-1}$ in acetonitrile as solvent. Figure 4.32 shows the results for $Eo = 4.4$; 6.8 and 9.8. Comparing with figure 4.30 reveals, that the hypothesis of same wake structures at constant Eötvös numbers is appropriate.

For Taylor bubbles with $Eo = 4.4$ a laminar wake is observed. Due to the low rise velocity of $w_b \approx 0.3 \ \text{cm·s}^{-1}$, the residence time is very long, so that the green colorization of the second reaction step is clearly observed. In case of $Eo = 6.8$ a turbulent wake structure results and no green colorization is observable due to the smaller residence time. With rising Eo, the wake structure becomes much more turbulent and the boundary layer is instable, so that a much higher mixing occurs. As already stated, this will result in different yield and selectivity in case of parallel/consecutive reactions.

It is likely, that the different residence times in case of Taylor bubbles lead to different mass transfer rates. Therefore, within the following experiment, two flow regimes, characterized by $Eo = 4.4$ and 5.5, with different reactant concentrations are investigated by characterization of the mass transfer coefficient.

The experimental procedure is similar to the one described at the beginning of this section (4.4.1). At first the mass transfer of oxygen in acetonitrile is determined as a reference. Therefore, the channel is filled with argon saturated acetonitrile and an oxygen bubble is injected. The bubble is then fixed in a counter current flow, so that the dissolution is observed over time. From the decrease of the equivalent diameter $d_{eq,V}$ with equation 4.26 the mass transfer coefficient is evaluated. Figure 4.33 shows the

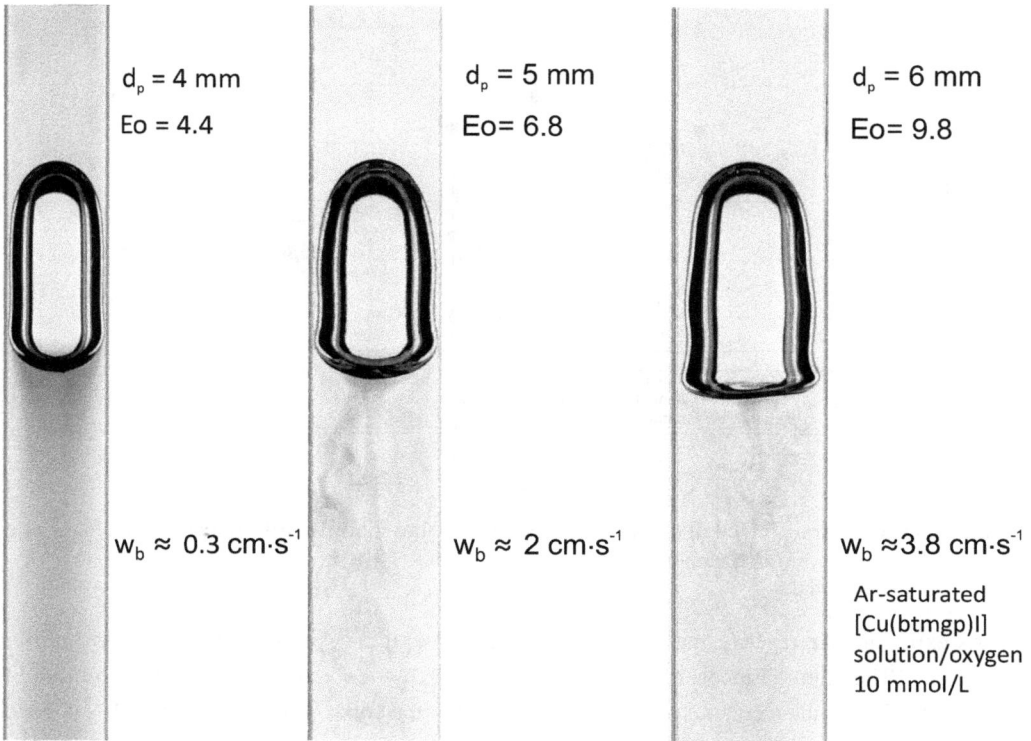

Figure 4.32 Backlit image of wake structures for the system Cu(btmgp)I-oxygen with a concentration of 10 mmol·L^{-1} at $Eo = 4.4; 6.8$ and 9.8 in argon saturated acetonitrile.

determined mass transfer coefficient for acetonitrile, in the following referred to as 0 mmol·L^{-1}, in dependency of the bubble equivalent diameter. The same investigations are performed for the concentrations of 1 and 10 mmol·L^{-1} Cu(btmgp)I. The mean mass transfer coefficients, the evaluated Hatta and Eötvös number are summarized in table 4.3.

Table 4.3 Listing of k_L^0, k_L^R, Ha, E, and E^* for the reaction of Cu(btmgp)I with oxygen in a Taylor flow capillary (4.5 mm diameter)

	result	value	unit	equation
mass transfer coefficient	$\overline{k_L^0}$	$3.10 \cdot 10^{-4} \pm 5.4 \cdot 10^{-5}$	m·s^{-1}	4.26
Hatta number	$\overline{Ha}$	$0.53 \pm 6.7 \cdot 10^{-2}$	-	2.32
theoretical enhancement factor	$\overline{E}$	$1.09 \pm 3.49 \cdot 10^{-2}$	-	2.33
reactive mass transfer coefficient	$\overline{k_{L,1}^R}$	$3.66 \cdot 10^{-4} \pm 5.2 \cdot 10^{-5}$	m·s^{-1}	4.26
experimental enhancement factor	$\overline{E_1^*}$	$1.18 \pm 1.9 \cdot 10^{-1}$	-	2.28
reactive mass transfer coefficient	$\overline{k_{L,10}^R}$	$5.34 \cdot 10^{-4} \pm 5.1 \cdot 10^{-5}$	m·s^{-1}	4.26
experimental enhancement factor	$\overline{E_{10}^*}$	$1.74 \pm 5.1 \cdot 10^{-1}$	-	2.28

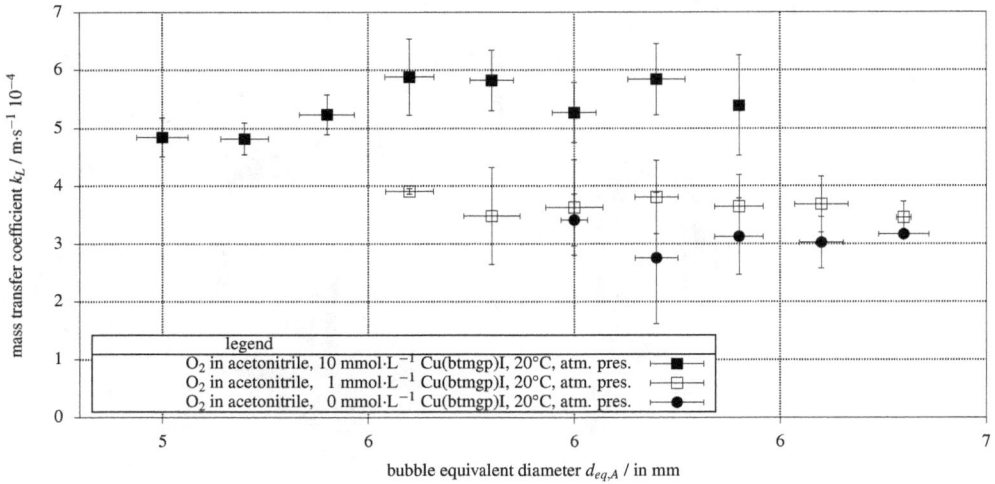

Figure 4.33 Mass transfer coefficient in dependency of the Cu(btmgp)I concentration (0, 1 and 10 mmol·L^{-1}) in a 4.5 mm channel ($Eo = 5.5$).

As expected, the reactive mass transfer coefficient is higher than the physical mass transfer coefficient. The Hatta number 0.53 is in the range of $0.3 < Ha < 3$ and with a reaction rate of 10^{-1} [Sch16], an enhancement factor of 1.09 results. For the concentration of 1 mmol·L^{-1} a good agreement of the E and E^* (deviation 8%) is observed. In case of a concentration of 10 mmol·L^{-1} the deviation is 60% and therefore much higher as predicted by the literature [Ast67].

For the channel diameter of 4 mm, respectively $Eo = 4.4$, a smaller mass transfer coefficient k_L^0 of $1.92 \cdot 10^{-4} \pm 4.8 \cdot 10^{-5}$ m·s^{-1} is determined, which agrees well with the assumed influence of the flow regime. The Hatta number is due to the lower mass transfer coefficient higher with a value of 0.86, which also results in an higher enhancement factor $E = 1.23$. The deviation of E and E^* for 1 and 10 mmol·L^{-1} is 15% and 12%.

Table 4.4 Listing of k_L^0, k_L^R, Ha, E, and E^* for the reaction of Cu(btmgp)I with oxygen in a Taylor flow capillary (4 mm diameter)

	result	value	unit	equation
mass transfer coefficient	$\overline{k_L^0}$	$1.92 \cdot 10^{-4} \pm 4.8 \cdot 10^{-5}$	m·s^{-1}	4.26
Hatta number	$\overline{Ha}$	$0.86 \pm 2.8 \cdot 10^{-2}$	-	2.32
theoretical enhancement factor	$\overline{E}$	$1.23 \pm 3.49 \cdot 10^{-2}$	-	2.33
reactive mass transfer coefficient	$\overline{k_{L,1}^R}$	$2.00 \cdot 10^{-4} \pm 8.8 \cdot 10^{-5}$	m·s^{-1}	4.26
experimental enhancement factor	$\overline{E_1^*}$	$1.04 \pm 2.1 \cdot 10^{-1}$	-	2.28
reactive mass transfer coefficient	$\overline{k_{L,10}^R}$	$2.67 \cdot 10^{-4} \pm 7.8 \cdot 10^{-5}$	m·s^{-1}	4.26
experimental enhancement factor	$\overline{E_{10}^*}$	$1.38 \pm 2.8 \cdot 10^{-1}$	-	2.28

As expected, the flow regime influences the mass transfer in case of Taylor bubbles. The different enhancement of mass transfer through the reaction is on one hand ascribed to inaccuracies in the measurement technique e.g. through injection of the gas phase. On the other hand, the assumption of a reaction kinetic of first order through an excess of oxygen, as used for the determination of the reaction rate, is probably incorrect.

For a detailed understanding of the flow regime influence on mass transfer and yield/selec-tivity of chemical reactions, local investigations of the concentration field are required. The fluorescence signal of the Cu(btmgp)I complex, as observed in preliminary investigations, is too low for a detection in p-LIF experiments. Therefore, the ligand system is adjusted by the group of Prof. Herres-Pawlis to enable a brighter fluorescence response. The result is the Cu(TMG$_2$tol)I complex, which shows a ten times higher fluorescence signal than the Cu(btmgp)I complex [Str17]. There are two major advantages of such a reactant. The first advantage is, that the reactant is also the fluorophore, so that no additional dye is added as an indicator for one of the reactants (e.g. oxygen). The second advantage is, that the calibration method of a serial dilution is possible to obtain concentration field data easily. Excitation of this fluorophore/reactant is only possible below 410 nm and the emission maximum is at 490 nm. In difference to the other investigations described within this work, a UV-Laser with a wavelength of 355 nm, 10 Hz repetition rate and an energy output of 80 mJ per pulse is used.

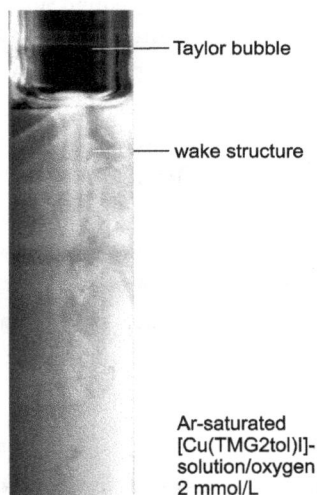

Figure 4.34 p-LIF image of the wake structures of the system Cu(TMG$_2$tol)I-oxygen with a concentra-tion of 2 mmol·L^{-1}, $d_p = 6$ mm, $Eo = 9.8$ in argon saturated acetonitrile.

Figure 4.34 shows the obtained gray level concentration field for an oxygen bubble rising in a solution of Cu(TMG$_2$tol)I (2 mmol·L^{-1}). As expected, the wake structure of the rising oxygen bubble is clearly observable in a loss in fluorescence intensity through the reaction of Cu(TMG$_2$tol)I with oxygen. A drawback of this fluorophore/reactant becomes also obvious. The absorption is relatively high, so that only 50% of the channel thickness are illuminated satisfactorily and lead to a misinterpretation of concentration field data.

The use of an internal fluorophore, for determining the concentration field and for the direct deter-mination of the yield/selectivity in combination with another fluorophore, is a promising approach to gain deeper knowledge of the interdependency of mass transfer, mixing and chemical reaction in combination with the Taylor bubble approach. This point will be investigated within the SPP 1740 during the second funding period.

The local investigation of chemical reactions is a challenging task in bubbly flows and especially for parallel/consecutive networks. Therefore, the reactive guiding measure Taylor bubble is developed, which allows the detailed investigation of chemical reaction under controlled hydrodynamic conditions. It is proven, that in dependency of the channel diameter and therefore the Eötvös number, adjustable hydrodynamic conditions are achieved (published in cooperation [Kas17]). The adjustable hydrodynamic conditions are plausible since the rise velocity of the Taylor bubble determines directly the flow conditions behind the rising bubble.

It can be concluded, that with rising channel diameter and velocity the mass transfer performance is enhanced for the non-reactive and the reactive case, which is consistent with the investigations of Hosoda and Kastens [Hos14, Kas15]. The values of physical mass transfer k_L^0 are always smaller than the reactive mass transfer k_L^R as expected due to the mass transfer enhancement of the reaction. The deviations in the mass transfer coefficients are likely a result of the experimental procedure, from the injection of the oxygen bubble to the first investigation, two to four seconds of reaction time are already elapsed in dependency of the bubble rise velocity. This will result in a different bubble shrinkage velocity due to different concentration gradients.

The calculated Hatta number for the reaction of sodium sulfite and the oxidation of Cu(btmgp)I are in the region of an intermediate reaction rate between $0.02 < Ha < 3$ as expected and show, that the setup is suitable for this reaction rate regime. By comparing the enhancement factors E and E^* for sodium sulfite, a good comparability of the theoretical and the experimental factor as expected from the literature is likely. The deviation is 18% and therefore only slightly higher than expected from the literature (max. 8% [Ast67]). In case of the Cu(btmgp)I oxidation this deviation is as well smaller and much higher with 15% or up to 60%. Additionally, a high influence of the reactant concentration is found, which is an evidence for a changing reaction order in dependency of the reactant concentration.

The established guiding measure reactive Taylor bubble is an excellent tool for the detailed and reproducible long term investigation of chemical reactions concerning the influence of different wake structures on yield and selectivity and will be used within the framework of the SPP 1740 in the next funding period.

4.5 Influence of Boundary Layer Deformations at Free Rising Bubbles

The identification of characteristic structures like separation points and vortices in the wake of wobbling gas bubbles is performed with the TRS-LIF technique (section 3.5.1) and allows the reconstruction of a three-dimensional concentration field within the bubble wake. Therefore, heavy image processing is required, since the recording is performed with 15 400 fps over 46 slices with a spacing of 1.08 mm. A background correction as already described in section 4.2 is required and therefore applied (figure 4.35, 1st). Afterwards a gray level spreading is performed to identify the wake structure (2nd) with a following band pass filtering and binarization is used to identify the wakes structure contour (3rd). The band pass filtering is applied since the laser sheet is inhomogeneous in vertical direction through defects on the polygon which propagate by deflection.

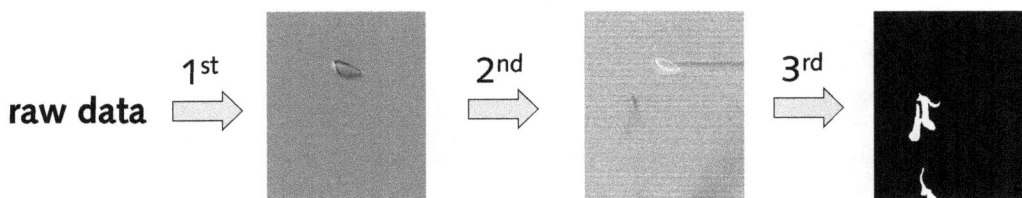

Figure 4.35 Image processing steps for TRS-LIF. 1st background correction; 2nd grey level spreading; 3rd band pass filtering and binarization.

Figure 4.36 shows a set of 11 images with an identified wake structure for a bubble with a diameter of $d_{eq,V}$ = 4.3 mm and a rise velocity of w_b = 26 cm·s^{-1}. Since 0.71 ms elapse during the recording of these 11 images, a motion blur of 3 pixel occurs. This small distortion is not corrected within the following. On the recording, the complex wake structure with a meandering structure and several separation points are observed. Additionally, the spatial extent and evolution of the detached vortices (e.g. figure 4.36 4-6) can be noticed.

To reconstruct the three-dimensional concentration wake and structure, a MATLAB$^{®}$ routine is developed. With this routine the recording is at first sorted with respect to the picture position. Therefore, the first image contains a small defect, a shadow from a PVC rod, for an automatic identification. Afterwards the pictures are processed as already described above. Now the pictures are separated again in recording sequences. Therefore, a time step, with in this case with 46 pictures, is created and the pictures are recombined according to this sequence. After sequencing the pictures, a three-dimensional reconstruction of the wake structure is performed by use of wake structure contour. The recognized and binarized structure is plotted in a three-dimensional coordinate system while the depth of the recording is expanded to the spacing of the light sheet stack (1.08 mm or 21 pixel). With this technique the three-dimensional wake structure (see figure 4.37) within the contour as boundary is reconstructed. Is is observed, that due to wobbling, the wake is detached irregularly and is representing the stochastic bubble deformation.

Additional to the wake contour also the concentration field is calculated (Figure 4.38). Therefore, the calibration pictures are processed similar to the experimental data and were sorted according to their spacial position. Afterwards for each plane a separate calibration equation is computed from the averaged data. The calibration is then applied to the matching plane to obtain the concentration field data. For a better visualization, the displayed concentration field is limited by the recognized contour of the wake and additionally an ellipsoid with the size of $d_{eq,V}$ = 4.3 mm is plotted at the identified bubble center. The size of the bubble is only an approximation from the bubble orientation and the size of the bubble shadow, since the exact bubble interface can not be determined within the LIF investigations, so that additional measuring techniques have to be applied in the future, that allow an exact distinction between the liquid and gas phase.

TRS-LIF is so far the only technique, that allows the determination of three-dimensional wake structures by obtaining simultaneously concentration field information. The gained information shows a very complex detachment and evolution of the wake structure with a different degree of mixing and dissolution. Besides areas with stable vortices, also areas of decamped wake structure are observable, which will lead to a varying yield and selectivity for parallel/consecutive reactions within the wake.

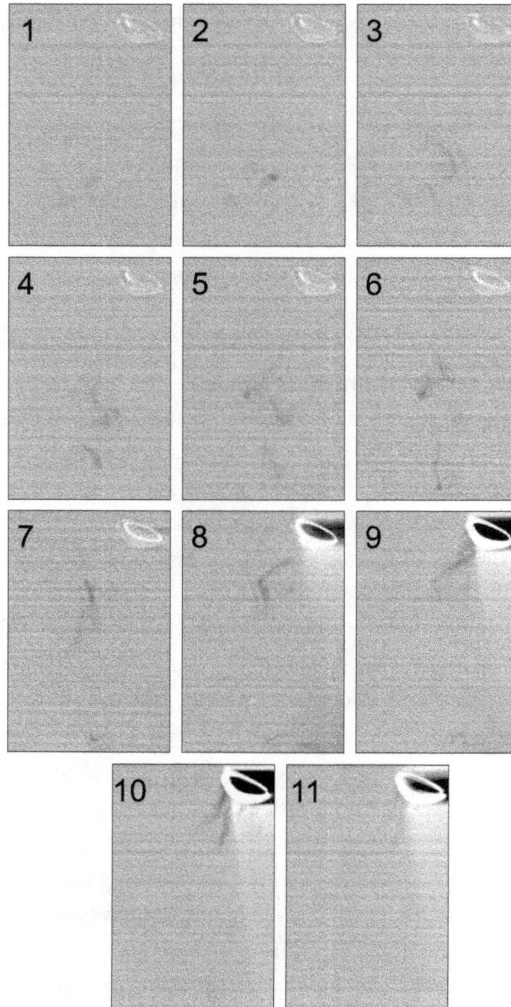

Figure 4.36 Set of 11 images with an identified wake structure for an ellipsoidal oxygen bubble ($d_{eq,V}$ = 4.3 mm; w_b =26 cm·s^{-1}).

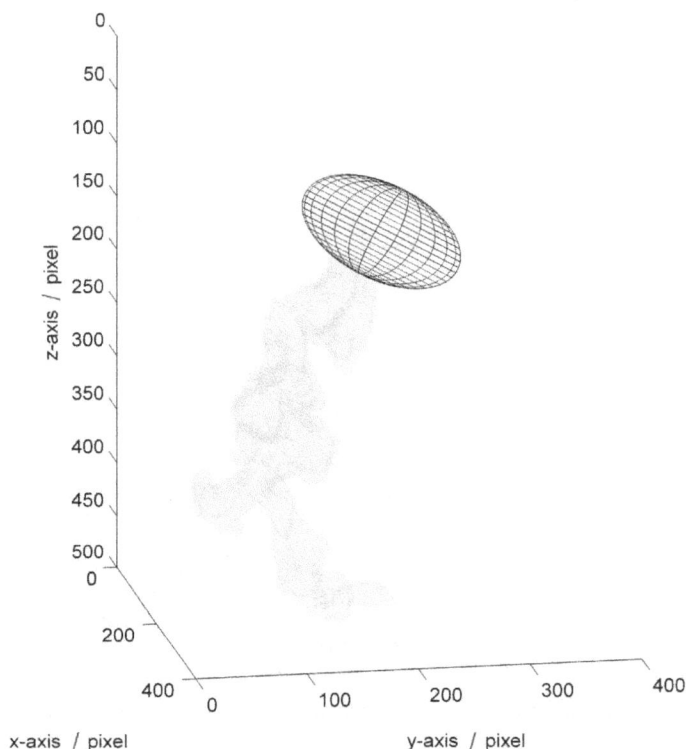

Figure 4.37 Reconstructed instantaneous wake structure of an oxygen bubble as three-dimensional iso-surface plot.

The investigation of irregular wake structures behind rising bubbles is a challenging task, since the rise velocity of the bubble with approx. $30 \ cm \cdot s^{-1}$ leads to very short observation times. Common techniques like p-LIF only partially reveal the wake structure of these bubbles due to a fixed light sheet position. Therefore, within this work the TRS-LIF technique (compare [Soo12]) was further developed and applied to rising bubbles for the first time. This technique allows the reconstruction of wake structures and reveals separation points as well as vortices with high reactant concentration.

The qualitative data of the wake structure show the capability of this method. The obtained structure is comparable to the results of Huang and Saito [Hua17] who investigated rising bubbles with p-LIF and changing light sheet position within several repeated experiments.

Within the future this technique will reveal the influence of mass transfer on the selectivity and yield of parallel/consecutive reactions.

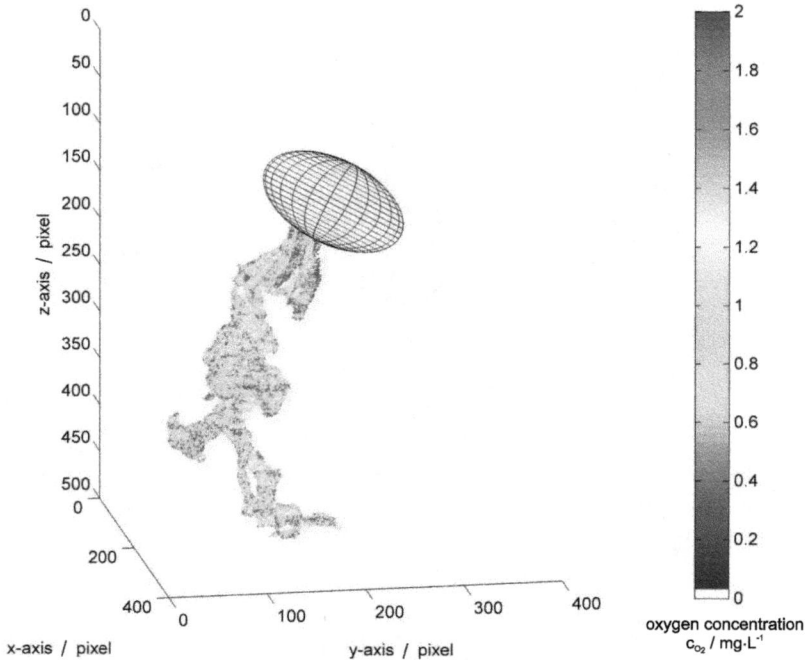

Figure 4.38 Reconstructed instantaneous wake structure of an oxygen bubble as three-dimensional concentration field.

4.6 Measurement Uncertainties and Problems

As for all measurements and techniques, uncertainties and problems occur, that have to be overcome for a reliable interpretation of the obtained data. Within the different LIF techniques (CLSM, p-LIF and TRS-LIF) used within this work, some very similar uncertainties occur and will be explained within the following.

One major error results from the supplied oxygen concentration for calibration and during the measurement. Since an equilibrium between gaseous and solved oxygen exists, small deviations within the experimental conditions, like temperature or pressure, lead to a varying oxygen concentration. While this concentration is monitored with an oxygen probe in each measurement, the deviations can't be fully compensated. Furthermore, the fluorophore shows a dependency of the pH value, which can't be considered within the calibration procedure, since the pH value is changing during the reaction progression. In case of reactive measurements this leads to the highest error, due to a change of the pH value from 6.7 to 9.3. A compensation of this error is not easily possible, since the pH value is also changing with reaction progression and a buffer will act as an inhibitor.

A further influence on the results inhere the evaluation methods itself and will therefore separately discussed.

4.6.1 CLSM

The intrinsic kinetics of the oxidation of sodium sulfite in an aqueous solution is determined by combination of experimental investigations and numerical simulations. A good agreement over a wide range of recorded concentration fields along the mixing chamber, as well as, in channel depth is obtained for the numerical simulation (see figure 4.2 and 4.3) in case of physical mixing. It is likely, that the experimental investigations and the numerical simulations both show an error that leads to the deviation. Since the numerical investigations are performed at the TU Dortmund, the possible errors are only briefly examined. First of all, the numerical investigations solve the CDR equation (equ. 2.40) based on the experimental parameters and geometry. The geometry is represented by the idealized geometry without any defects and surface roughness. While the surface roughness can be implemented within the numerical simulation, the determination and implementation of defects, which result from the lithography process for channel production, is nearly impossible. Not all defects are detected by a nondestructive investigation, so that there are always deviations between experimental and numerical investigations. Errors within the numerical solvers, like numerical diffusion are suppressed, but can not be fully excluded. The investigation of these errors is not a topic of this work and will therefore not be further discussed.

The numerical simulation depends mainly on the experimental conditions like flow rate, concentrations and geometry. While the error of the geometry is tough to grasp, the experimental error for the flow rate is mainly depending on the accurateness of the used mass flow controllers. The maximal error is $\pm$ 12.5 g·h^{-1} ($\pm$ 0.5% of the maximal flow rate) according to the manufacturer and is not reduced through a calibration. Therefore, the error at very small flow rates, like 15 g·h^{-1} within this work, is nearly 100% of the set value. This high error explains the deviation of the diffusion gradient position between measurement and simplified CDR model in figure 4.12 c). Furthermore, the start concentration influences the comparability to the results. Through the pressure induced flow, an over saturation of oxygen results, that is measured with an oxygen probe and is for all measurements in the range of 12 $\pm$0.5 mg·L^{-1}, so that small deviations in the start conditions exist, but this deviation is not influencing the diffusion gradient position.

4.6.2 p-LIF

In case of the p-LIF investigations, the recorded image is first of all background corrected, than processed to determine the bubble center. Afterwards the recording is cropped according to the bubble position and a mean image of all resulting images is calculated. Within this post processing procedure errors that are caused through the laser light sheet, camera and lenses are compensated, so an overall smaller error results. Nevertheless, all post processing corrections have drawbacks and imperfections, so that through the applied correction a blur of the bubble and the bubble wake occurs, but yet an evaluation as applied within this work will fail without the superposition of several images since high speed p-LIF shows a drawback that is in common p-LIF already overcome. High background noise exists due to a insensitive camera CMOS sensor.

Additionally, to the blur of the first post processing procedure, the allocation of concentrations to the observed gray levels by use of the calibration equation leads to a further error. The background noise is already suppressed by the background correction and leads only to a small error, but there are still small brightness differences within the image that will be misinterpreted by the algorithm and lead to errors of 0.05 mg·L^{-1}.

With the evaluated concentration fields the application of a mass balance for this symmetric wake structure is easily possible. Therefore, all data below 0.1 mg·L^{-1} are excluded to determine the radius of the bubble wake. Through the blur and brightness changes an error of 1 to 2 pixel occurs, which is affecting the determined mass transfer coefficient by a value of 0.1·10^{-4} m·s^{-1}.

For the determination of the rise velocity and bubble size it is assumed, that the bubble shadow corresponds to the size of a spherical bubble and the velocity of the shadow front is equal to the velocity of the bubble center. As described, a MATLAB$^{®}$ routine is used, which determines the shadow size and front by setting a threshold. With this method a systematic error results. While a laser sheet correction is performed to minimize the non uniform illumination of the LIF background, the correction is not perfect, so that gray level deviations of up to 10 values exist. This can lead to an error within the determination of shadow size and velocity of 2 pixel, which results in an error in size of 0.01 mm and 5 mm·s^{-1} in velocity.

It has to be stated, that the results are still representing only a very short time period ($\approx$ 0.05 s) which is averaged, so that the determined mass transfer coefficient is only a snapshot of the overall process and fluctuations can therefore not be compensated. Changes in mass transfer performance, as found by [Noc16], due to a change in the rising mode (zig-zag to rectilinear) are unlikely, since only a rectilinear rise is observed. Nevertheless, small deviations of the bubble rise position through the bubble generation process are likely and explain the observed variance.

4.6.3   TRS-LIF

There are further improvements of this technique possible. Through adjustment of the polygon diameter and number of mirrors, the spacing between the LIF planes is narrowed, to reach a more detailed reconstructed wake structure. Additionally, this is achieved by a higher repetition rate of the laser beam, but therefore also a higher recording rate of the camera is required, which will lead to a poor image quality. Nevertheless, the recording camera is the bottleneck so far, since the high recording rates are achieved by cropping the image size, which leads to a small resolution. Additionally, the sensitivity of the camera sensor determines the signal-to-noise ratio of the recorded image, so that a more sensitive sensor will improve the image quality significantly.

Chapter 5

Overall Discussion

The approach of separating the timescales of mixing, mass transfer and reaction has proven to be suitable for gaining a deeper insight and understanding of the processes in reactive bubbly flows. Among other things, several key informations for modeling and numerical simulations are achieved. By use of CLSM investigations within the SFM combined with numerical simulations the intrinsic kinetics of the sodium sulfite reaction is measured. Additionally, a simplified CDR model is developed, which allows estimation of the kinetics without time intensive numerical simulations.

Based on the rise velocity of rectilinear rising bubbles, the influence of the fluorophore, Dichlorotris (1,10-phenanthroline)ruthenium(II)hydrate, is proven. In addition, the mass transfer coefficient is determined by means of p-LIF and quantified with a mass balance. The results agree well with the mass transfer correlation of Frössling [Fro38] combined with the rise velocity prediction for contaminated bubbles of Tomiyama [Tom98] and proves again the surfactant behavior of the fluorophore.

Even though, a suitable parallel/consecutive reaction system is not available so far, the investigations of bouncing bubbles and free rising bubbles shows, that the detachment of the boundary layer and induced vortices can lead to a local affectation of the yield and selectivity. The adapted TRS-LIF technique provides new insights in the mass transfer processes and wake structure evolution of the resulting wake structure.

With the developed and sharpened methods presented within this work, an easy and reliable investigation on parallel/consecutive reactions is possible and will lead to a better understanding of such processes. The proposed approach for a suitable parallel/consecutive reaction system with reactant/fluorophore character is:

1st Determination of the intrinsic kinetics by the use of SuperFocus-Mixer as required parameter for numerical investigations.

2nd p-LIF at rectilinear rising bubbles for comparison and validation of numerical with experimental concentration fields. The results serve as a first reference to understand the local yield and selectivity at the boundary layer as proposed by [Khi01, Khi03, Fal17]

3rd Overall mass transfer and p-LIF investigations at Taylor bubbles for a detailed understanding of the influence of mixing and wake structure on the local yield and selectivity.

4th TRS-LIF investigations at free rising bubbles with shape deformations to gain knowledge about the influence on the yield and selectivity in a bubbly flow.

5th Transfer to industrial bubbly flows and bubble columns by the use of models and numerical simulations.

The newly applied TRS-LIF technique already reveals the three-dimensional wake structure, but the lateral resolution in depth needs to be improved to allow a detailed investigation with mass balance for quantification of the transferred mass. A higher resolution with the current experimental setup is impossible due to the fixed polygon size and mirrors. With a higher rotation speed of a smaller polygon, the spacing within the light sheet stack can be lowered, so that a higher lateral resolution will result.

As soon as a suitable parallel/consecutive reaction system is available, the above proposed approach will reveal the influence of bouncing and shape deformations on the yield and selectivity. Once this information are available, a major improvement in the description, prediction and certainty of bubble columns in industrial processes will occur, so that an energy and resource minimum is reached.

Chapter 6

Conclusion

The benchmark facilities developed within this work prove, that an influence of mixing on mass transfer exists. Based on the experimental findings it can be concluded, that mass transfer is enhanced on a micro scale by bubble bouncing and due to different wake structures. In consequence of the dependency on macro scale for yield and selectivity, it is likely, that the found influence on micro scale will affect parallel/consecutive reactions. Furthermore, it is proven, that the used fluorophore Dichlorotris-(1,10-phenanthroline)-ruthenium(II)-hydrate have to be considered as surface active and all evaluated data are only valid for a contaminated watery system.

In addition, new insights in the processes at bubble-bubble interactions (bouncing) are achieved with the p-LIF technique. It is shown, that bubble bouncing leads to an intensified mass transfer through detachment of the bubble wake and the development of vortices with higher concentration. Additionally, the level of mass transfer enhancement is determined for different collision frequencies and a model is developed. This model can be applied to numerical investigations of bubbly flows to describe the influence of bouncing, which is so far not considered in these investigations.

To overcome the influence of surface active agents and to enable detailed investigations on the influence of mixing within wake structures on chemical reactions, Taylor bubbles in organic solvents have proven as an excellent tool in combination with the reaction systems developed within the SPP 1740 "Reactive Bubbly Flows". Therefore, the work of Kastens *et al.* [Kas15] is extended and transferred to investigations on reactive mass transfer. It turned out, that Taylor bubble investigations enable very detailed investigations on the interdependence of mixing and mass transfer and will be easily applied to parallel/consecutive reactions, once a suitable reaction system is developed. For the investigation of free rising bubbles, the method of TRS-LIF is further developed and successful used to visualize the three-dimensional wake structure behind rising bubbles.

Finally it should be noted, that within this work the fundamental procedure and tools for the clarification of boundary layer deformations and their influence on mass transfer are developed and successfully used. Although the influence on yield and selectivity for a parallel/consecutive reaction is not finally resolved, the developed procedure and tools will enable detailed investigations within the SPP 1740 framework in the near future.

Bibliography

[Abe08] Abe, S., Okawa, H., Hosokawa, S. and Tomiyama, A. *Dissolution of a carbon dioxide bubble in a vertical pipe.* Journal of Fluid Science and Technology, 3(5):667–677, 2008.

[Alp83] Alper, E. *Mass Transfer with Chemical Reaction in Multiphase Systems.* Martinus Nijhoff Publishers, 1983.

[Álv00] Álvarez, E., Sanjurjo, B., Cancela, A. and Navaza, J. *Mass transfer and influence of physical properties of solutions in a bubble column.* Chemical Engineering Research and Design, 78(6):889–893, 2000.

[Alv04] Alves, S., Maia, C. and Vasconcelos, J. *Gas-liquid mass transfer coefficient in stirred tanks interpreted through bubble contamination kinetics.* Chemical Engineering and Processing: Process Intensification, 43(7):823–830, 2004.

[Alv05] Alves, S., Orvalho, S. and Vasconcelos, J. *Effect of bubble contamination on rise velocity and mass transfer.* Chemical Engineering Science, 60(1):1–9, 2005.

[And67] Anderson, R.A. *Fundamentals of vibrations.* Macmillan, 1967.

[Aok15] Aoki, J., Hayashi, K. and Tomiyama, A. *Mass transfer from single carbon dioxide bubbles in contaminated water in a vertical pipe.* International Journal of Heat and Mass Transfer, 83:652–658, 2015.

[Ast67] Astarita, G. *Mass transfer with chemical reaction.* Elsevier, 1967.

[Atk06] Atkins, P.W. and De Paula, J. *Physikalische Chemie*, vol. 4. John Wiley & Sons, 2006.

[Bäc27] Bäckström, H.L. *The chain-reaction theory of negative catalysis1.* Journal of the American Chemical Society, 49(6):1460–1472, 1927.

[Bäc34] Bäckström, H.L. *Der Kettenmechanismus bei der Autoxydation von Aldehyden.* Zeitschrift für physikalische Chemie, 25(1):99–121, 1934.

[Bae87] Baerns, M., Hofmann, H. and Renken, A. *Chemische Reaktionstechnik.* Thieme Stuttgart, 1987.

[Bał99] Bałdyga, J. and Bourne, J.R. *Turbulent mixing and chemical reactions.* Wiley, 1999.

[Bar03] Baroud, C.N., Okkels, F., Ménétrier, L. and Tabeling, P. *Reaction-diffusion dynamics: Confrontation between theory and experiment in a microfluidic reactor.* Physical Review E, 67(6):060104, 2003.

[Ben65] Bennett, A., Hewitt, G., Kearsey, H., Keeys, R. and Lacey, P. *Flow visualisation studies of boiling at high pressure.* Proc. Inst. Mech. Eng., 180:1–11, 1965.

[Bis91] Bischof, F., Sommerfeld, M. and Durst, F. *The determination of mass transfer rates from individual small bubbles.* Chemical engineering science, 46(12):3115–3121, 1991.

[Bou03] Bourne, J.R. *Mixing and the selectivity of chemical reactions. Organic Process Research & Development*, 7(4):471–508, 2003.

[Bra71a] Brauer, H. *Grundlagen der Einphasen-und Mehrphasenströmungen*, vol. 2. Sauerländer, 1971.

[Bra71b] Brauer, H. and Mewes, D. *Stoffaustausch einschliesslich chemischer Reaktionen*. Verlag Sauerlander, 1971.

[Brü99] Brücker, C. *Structure and dynamics of the wake of bubbles and its relevance for bubble interaction. Physics of fluids*, 11(7):1781–1796, 1999.

[Cal61] Calderbank, P. and Moo-Young, M. *The continuous phase heat and mass-transfer properties of dispersions. chemical Engineering science*, 16(1-2):39–54, 1961.

[Cli71] Clift, R. and Gauvin, W. *Motion of particles in turbulent gas streams. British Chemical Engineering*, 16(2-3):229, 1971.

[Cli78] Clift, R., Grace, J. and Weber, M. *Bubbles, drops, and particles*. Dover Publ., 1978.

[Com71] Comolet, R. *Ascension rate of gas bubbles in a slightly viscous liquid. Comptes rendus hebdomadaires des seances de l academie des science serie A*, 272(18):1213, 1971.

[Com17] Compart, C. *Design of an Experimental Setup for the Analysis of Mass Transfer on Reactive Taylor-Bubbles*. master thesis, 2017.

[Con95] Connick, R.E., Zhang, Y.X., Lee, S., Adamic, R. and Chieng, P. *Kinetics and mechanism of the oxidation of HSO3-by O2. 1. the uncatalyzed reaction. Inorganic Chemistry*, 34(18):4543–4553, 1995.

[Cra79] Crank, J. *The mathematics of diffusion*. Oxford university press, 1979.

[Cri08] Crimaldi, J. *Planar laser induced fluorescence in aqueous flows. Experiments in fluids*, 44(6):851–863, 2008.

[Dan51] Danckwerts, P. *Significance of liquid-film coefficients in gas absorption. Industrial & Engineering Chemistry*, 43(6):1460–1467, 1951.

[Dan07] Dani, A., Guiraud, P. and Cockx, A. *Local measurement of oxygen transfer around a single bubble by planar laser-induced fluorescence. Chemical Engineering Science*, 62(24):7245–7252, 2007.

[Dav50] Davies, R. and Taylor, G. *The mechanics of large bubbles rising through extended liquids and through liquids in tubes. Proceedings of the Royal Society of London. Series A, Mathematical and Physical Sciences*, pp. 375–390, 1950.

[Dav57] Davidson, J. and Cullen, E. *The determination of diffusion coefficients for sparingly soluble gases in liquids. Trans. Inst. Chem. Eng*, 35:51–60, 1957.

[Deu01] Deusch, S. and Dracos, T. *Time resolved 3D passive scalar concentration-field imaging by laser induced fluorescence (LIF) in moving liquids. Measurement Science and Technology*, 12(2):188, 2001.

[Dog67] Dogliotti, L. and Hayon, E. *Flash photolysis of per [oxydi] sulfate ions in aqueous solutions. The sulfate and ozonide radical anions. The Journal of Physical Chemistry*, 71(8):2511–2516, 1967.

[Dre04] Drese, K.S. *Optimization of interdigital micromixers via analytical modeling—exemplified with the SuperFocus mixer. Chemical Engineering Journal*, 101(1):403–407, 2004.

[Dui94] Duineveld, P.C. *Bouncing and coalescence of two bubbles in water*. Ph. D. thesis, Twente University, 1994.

[Dui95] Duineveld, P. *The rise velocity and shape of bubbles in pure water at high Reynolds number.* Journal of Fluid Mechanics, 292:325–332, 1995.

[Duk15] Dukhin, S., Kovalchuk, V., Gochev, G., Lotfi, M., Krzan, M., Malysa, K. and Miller, R. *Dynamics of Rear Stagnant Cap formation at the surface of spherical bubbles rising in surfactant solutions at large Reynolds numbers under conditions of small Marangoni number and slow sorption kinetics.* Advances in colloid and interface science, 222:260–274, 2015.

[Dum43] Dumitrescu, D.T. *Strömung an einer Luftblase im senkrechten Rohr.* ZAMM-Journal of Applied Mathematics and Mechanics/Zeitschrift für Angewandte Mathematik und Mechanik, 23(3):139–149, 1943.

[Erm01] Ermakov, A. *et al.. Catalytic mechanism of the "noncatalytic" autooxidation of sulfite.* Kinetics and catalysis, 42(4):479–489, 2001.

[Erm02] Ermakov, A. *et al.. Catalysis of HSO3–/SO32–Oxidation by Manganese Ions.* Kinetics and Catalysis, 43(2):249–260, 2002.

[Fal17] Falcone, M., Dieter, B. and Holger, M. *3D Direct numerical simulations of reactive mass transfer from deformable single bubbles: an analysis of mass transfer coefficients and reaction selectivities.* Chemical Engineering Science, 2017. ISSN 0009-2509. doi: https://doi.org/10.1016/j.ces.2017.11.024.

[Fra31] Franck, J. and Haber, F. *Zur Theorie der Katalyse durch Schwermetallionen in wässriger Lösung und insbesondere zur Autoxydation der Sulfitlösungen.* Verlag d. Akad. d. Wissenschaften, de Gruyter, 1931.

[Fro38] Froessling, N. *The evaporation of falling drops (in German).* Gerlands Beiträge zur Geophysik, 52:170–216, 1938.

[Gas15] Gaspar, J. and Fosbøl, P.L. *A general enhancement factor model for absorption and desorption systems: a CO 2 capture case-study.* Chemical Engineering Science, 138:203–215, 2015.

[Gla77] Glaeser, H. and Brauer, H. *Berechnung des Impuls-und Stofftransports durch die Grenzfläche einer formveränderlichen Blase.* VDI-Verlag, 1977.

[Gri62] Griffith, R. *The effect of surfactants on the terminal velocity of drops and bubbles.* Chemical Engineering Science, 17(12):1057–1070, 1962.

[Had11] Hadamard, J. *Mouvement permanent lent d'une sphère liquide et visqueuse dans un liquide visqueux.* C. R. Acad. Sci., 152(1):1735–1752, 1911.

[Har60] Harmathy, T.Z. *Velocity of large drops and bubbles in media of infinite or restricted extent.* AIChE Journal, 6(2):281–288, 1960.

[Har03] Hardt, S. and Schönfeld, F. *Laminar mixing in different interdigital micromixers: II. Numerical simulations.* AIChE journal, 49(3):578–584, 2003.

[Has12] Hassanvand, A. and Hashemabadi, S.H. *Direct numerical simulation of mass transfer from Taylor bubble flow through a circular capillary.* International Journal of Heat and Mass Transfer, 55(21):5959–5971, 2012.

[Hat32] Hatta, S. *On the absorption velocity of gases by liquids.* Tech. Repts. Tohoku Imp. Univ, 10:119–128, 1932.

[Hay14] Hayashi, K., Hosoda, S., Tryggvason, G. and Tomiyama, A. *Effects of shape oscillation on mass transfer from a Taylor bubble.* International Journal of Multiphase Flow, 58:236–245, 2014.

[Her03] Hermann, R., Lehmann, M. and Büchs, J. *Characterization of gas–liquid mass transfer phenomena in microtiter plates. Biotechnology and Bioengineering*, 81(2):178–186, 2003.

[Hes03] Hessel, V., Hardt, S., Löwe, H. and Schönfeld, F. *Laminar mixing in different interdigital micromixers: I. Experimental characterization. AIChE Journal*, 49(3):566–577, 2003.

[Hig35] Higbie, R. *The rate of absorption of a pure gas into a still liquid during short periods of exposure. Trans. AIChE*, 31:365–389, 1935.

[Hik64] Hikita, H. and Asai, S. *Gas absorption with (m, n)-th order irreversible chemical reaction. Int. Chem. Eng*, 4(2):332–340, 1964.

[Hla17] Hlawitschka, M.W., Oßberger, M., Backes, C., Klüfers, P. and Bart, H.J. *Reactive Mass Transfer of Single NO Bubbles and Bubble Bouncing in Aqueous Ferric Solutions–A Feasibility Study. Oil & Gas Science and Technology–Revue d'IFP Energies nouvelles*, 72(2):11, 2017.

[Hos14] Hosoda, S., Abe, S., Hosokawa, S. and Tomiyama, A. *Mass transfer from a bubble in a vertical pipe. International Journal of Heat and Mass Transfer*, 69:215–222, 2014.

[Hua17] Huang, J. and Saito, T. *Influences of gas–liquid interface contamination on bubble motions, bubble wakes, and instantaneous mass transfer. Chemical Engineering Science*, 157:182–199, 2017.

[Jia17] Jia, H. and Zhang, P. *Mass transfer of a rising spherical bubble in the contaminated solution with chemical reaction and volume change. International Journal of Heat and Mass Transfer*, 110:43–57, 2017.

[Jim13] Jimenez, M., Dietrich, N. and Hébrard, G. *Mass transfer in the wake of non-spherical air bubbles quantified by quenching of fluorescence. Chemical Engineering Science*, 100:160–171, 2013.

[Kar08] Karatza, D., Prisciandaro, M., Lancia, A. and Musmarra, D. *Reaction rate of sulfite oxidation catalyzed by cuprous ions. Chemical Engineering Journal*, 145(2):285–289, 2008.

[Kaš82] Kaštánek, F. and Fialová, M. *Possible application of approximate models for calculation of selectivity of consecutive reactions under the effect of mass transfer. Collection of Czechoslovak Chemical Communications*, 47(5):1301–1309, 1982.

[Kaš93] Kaštánek, F., Zahradník, J., Kratochvíl, J. and Cermák, J. *Chemical reactors for gas-liquid systems*. Ellis Horwood, 1993.

[Kas14] Kashid, M.N., Renken, A. and Kiwi-Minsker, L. *Microstructured devices for chemical processing*. John Wiley & Sons, 2014.

[Kas15] Kastens, S., Hosoda, S., Schlüter, M. and Tomiyama, A. *Mass transfer from single taylor bubbles in minichannels. Chemical Engineering & Technology*, 38(11):1925–1932, 2015.

[Kas17] Kastens, S., Timmermann, J., Strassl, F., Rampmaier, R., Hoffmann, A., Herres-Pawlis, S. and Schlüter, M. *Test System for the Investigation of Reactive Taylor Bubbles. Chemical Engineering & Technology*, 40(8):1494–1501, 2017.

[Khi01] Khinast, J.G. *Impact of 2-D bubble dynamics on the selectivity of fast gas-liquid reactions. AIChE Journal*, 47(10):2304–2319, 2001.

[Khi03] Khinast, J.G., Koynov, A.A. and Leib, T.M. *Reactive mass transfer at gas–liquid interfaces: impact of micro-scale fluid dynamics on yield and selectivity of liquid-phase cyclohexane oxidation. Chemical Engineering Science*, 58(17):3961–3971, 2003.

[Koy04] Koynov, A. and Khinast, J.G. *Effects of hydrodynamics and Lagrangian transport on chemically reacting bubble flows.* Chemical Engineering Science, 59(18):3907–3927, 2004.

[Koy05] Koynov, A., Khinast, J.G. and Tryggvason, G. *Mass transfer and chemical reactions in bubble swarms with dynamic interfaces.* AIChE Journal, 51(10):2786–2800, 2005.

[Kra12] Kraume, M. *Transportvorgänge in der Verfahrenstechnik Grundlagen und apparative Umsetzungen, 2. bearb. Auflage.* Springer-Vieweg Wiesbaden, 2012.

[Küc09] Kück, U.D., Schlüter, M. and Räbiger, N. *Analyse des grenzschichtnahen Stofftransports an frei aufsteigenden Gasblasen.* Chemie Ingenieur Technik, 81(10):1599–1606, 2009.

[Küc10] Kück, U., Schlüter, M. and Räbiger, N. *Investigation on reactive mass transfer at freely rising gas bubbles experimental methods for multiphase flows.* In *Seventh International Conference on Multiphase Flow, Florida, USA.* 2010.

[Küc11] Kück, U.D., Kröger, M., Bothe, D., Räbiger, N., Schlüter, M. and Warnecke, H.J. *Skalenübergreifende Beschreibung der Transportprozesse bei Gas/Flüssig-Reaktionen.* Chemie Ingenieur Technik, 83(7):1084–1095, 2011.

[Kul05] Kulkarni, A.A. and Joshi, J.B. *Bubble formation and bubble rise velocity in gas- liquid systems: a review.* Industrial & Engineering Chemistry Research, 44(16):5873–5931, 2005.

[Kur13] Kurimoto, R., Hayashi, K. and Tomiyama, A. *Terminal velocities of clean and fully-contaminated drops in vertical pipes.* International Journal of Multiphase Flow, 49:8–23, 2013.

[Lai56] Laird, A. and Chisholm, D. *Pressure and forces along cylindrical bubbles in a vertical tube.* Industrial & Engineering Chemistry, 48(8):1361–1364, 1956.

[Lan97] Lancia, A., Musmarra, D., Pepe, F. and Prisciandaro, M. *Model of oxygen absorption into calcium sulfite solutions.* Chemical Engineering Journal, 66(2):123–129, 1997.

[Lea85] Leaist, D.G. *Moments analysis of restricted ternary diffusion: sodium sulfite+ sodium hydroxide+ water.* Canadian journal of chemistry, 63(11):2933–2939, 1985.

[Lev62] Levich, V.G. *Physicochemical hydrodynamics.* Prentice hall, 1962.

[Lev72] Levenspiel, O. *Chemical reaction engineering*, vol. 2. Wiley, 1972.

[Lev99] Levenspiel, O. *Chemical reaction engineering*, vol. 3. Wiley, 1999.

[Lew24] Lewis, W. and Whitman, W. *Principles of gas absorption.* Industrial & Engineering Chemistry, 16(12):1215–1220, 1924.

[Lin81] Linek, V. and Vacek, V. *Chemical engineering use of catalyzed sulfite oxidation kinetics for the determination of mass transfer characteristics of gas—liquid contactors.* Chemical Engineering Science, 36(11):1747–1768, 1981.

[Liu09] Liu, W. *Elementary feedback stabilization of the linear reaction-convection-diffusion equation and the wave equation*, vol. 66. Springer Science & Business Media, 2009.

[Liu15] Liu, D., Wall, T. and Stanger, R. *CO 2 quality control through scrubbing in oxy-fuel combustion: Rate limitation due to S (IV) oxidation in sodium solutions in scrubbers and prior to waste disposal.* International Journal of Greenhouse Gas Control, 39:148–157, 2015.

[Mad07] Madhavi, T., Golder, A., Samanta, A. and Ray, S. *Studies on bubble dynamics with mass transfer.* Chemical engineering journal, 128(2):95–104, 2007.

[Mar65] Marrucci, G. *Communication. Rising Velocity of Swarm of Spherical Bubbles. Industrial & engineering chemistry fundamentals*, 4(2):224–225, 1965.

[Men67] Mendelson, H.D. *The prediction of bubble terminal velocities from wave theory. AIChE Journal*, 13(2):250–253, 1967.

[Mer77] Mersmann, A. *Auslegung und Maßstabsvergrößerung von Blasen-und Tropfensäulen. Chemie Ingenieur Technik*, 49(9):679–691, 1977.

[Mie17] Mierka, O., Munir, M., Spille, C., Timmermann, J., Schlüter, M. and Turek, S. *Reactive Liquid-Flow Simulation of Micromixers Based on Grid Deformation Techniques. Chemical Engineering & Technology*, 2017.

[Mor16] Morgado, A., Miranda, J., Araújo, J. and Campos, J. *Review on vertical gas–liquid slug flow. International Journal of Multiphase Flow*, 85:348–368, 2016.

[Nic62] Nicklin, D. *Two-phase bubble flow. Chemical Engineering Science*, 17(9):693–702, 1962.

[Noc16] Nock, W., Heaven, S. and Banks, C. *Mass transfer and gas–liquid interface properties of single CO 2 bubbles rising in tap water. Chemical Engineering Science*, 140:171–178, 2016.

[Ohl01] Ohl, C. *Generator for single bubbles of controllable size. Review of Scientific Instruments*, 72(1):252–254, 2001.

[Orh16] Orhan, R. and Dursun, G. *Effects of surfactants on hydrodynamics and mass transfer in a co-current downflow contacting column. Chemical Engineering Research and Design*, 109:477–485, 2016.

[Pai05] Painmanakul, P., Loubière, K., Hébrard, G., Mietton-Peuchot, M. and Roustan, M. *Effect of surfactants on liquid-side mass transfer coefficients. Chemical Engineering Science*, 60(22):6480–6491, 2005.

[Pee53] Peebles, F.N. *Studies on the motion of gas bubbles in liquids. Chem. Eng. Prog.*, 49:88–97, 1953.

[Pet12] Peters, F. and Els, C. *An experimental study on slow and fast bubbles in tap water. Chemical engineering science*, 82:194–199, 2012.

[Räb13] Räbiger, N. and Schlüter, M. *Bildung und Bewegung von Tropfen und Blasen. VDI Wärmeatlas*, 2013.

[Roe01] Roessler, A. and Rys, P. *Wenn die Rührgeschwindigkeit die Produktverteilung bestimmt: Selektivität mischungsmaskierter Reaktionen. Chemie in unserer Zeit*, 35(5):314–323, 2001.

[Rol11] Rolff, M., Schottenheim, J., Decker, H. and Tuczek, F. *Copper–O 2 reactivity of tyrosinase models towards external monophenolic substrates: molecular mechanism and comparison with the enzyme. Chemical Society Reviews*, 40(7):4077–4098, 2011.

[Rud15] Rudnik, K. *Investigations on the influence of bubble collisions on mass transfer into the bulk phase.* master thesis, 2015.

[San09] Sanada, T., Sato, A., Shirota, M. and Watanabe, M. *Motion and coalescence of a pair of bubbles rising side by side. Chemical Engineering Science*, 64(11):2659–2671, 2009.

[Sar06] Sardeing, R., Painmanakul, P. and Hébrard, G. *Effect of surfactants on liquid-side mass transfer coefficients in gas–liquid systems: a first step to modeling. Chemical Engineering Science*, 61(19):6249–6260, 2006.

[Sch02] Schlüter, M. *Blasenbewegung in praxisrelevanten Zweiphasenströmungen.* VDI-Verlag, 2002.

[Sch16] Schurr, D., Strassl, F., Liebhäuser, P., Rinke, G., Dittmeyer, R. and Herres-Pawlis, S. *Decay kinetics of sensitive bioinorganic species in a SuperFocus mixer at ambient conditions. Reaction Chemistry & Engineering*, 1(5):485–493, 2016.

[Sem53] Semenov, N. and Bradley, J. *Some problems of chemical kinetics and reactivity.* Pergamon Press, 1953.

[Sha82] Shah, Y.T., Kelkar, B.G., Godbole, S.P. and Deckwer, W.D. *Design parameters estimations for bubble column reactors. AIChE Journal*, 28(3):353–379, 1982. ISSN 1547-5905. doi:10.1002/aic.690280302.

[She12] Shen, Z., Guo, S., Kang, W., Zeng, K., Yin, M., Tian, J. and Lu, J. *Kinetics and mechanism of sulfite oxidation in the magnesium-based wet flue gas desulfurization process. Industrial & Engineering Chemistry Research*, 51(11):4192–4198, 2012.

[Shu12] Shuai, C., Lidong, W., Siqi, H. and Leixia, D. *Oxidation rate of sodium sulfite in presence of inhibitors. Energy Procedia*, 16:2060–2066, 2012.

[Som13] Sommerfeld, M. *Bewegung fester Partikel in Gasen und Flüssigkeiten. VDI-Wärmeatlas.* Berlin: Springer, 2013.

[Son08] Sone, D., Sakakibara, K., Yamada, M., Sanada, T. and Saito, T. *Bubble motion and its surrounding liquid motion through the collision of a pair of bubbles. Journal of Power and Energy Systems*, 2(1):306–317, 2008.

[Soo12] Soodt, T., Schröder, F., Klaas, M., van Overbrüggen, T. and Schröder, W. *Experimental investigation of the transitional bronchial velocity distribution using stereo scanning PIV. Experiments in fluids*, 52(3):709–718, 2012.

[Spi15] Spille, C. *Determination of microkinetics by investigations on local concentration fields within a SuperFocus mixer.* project theses, 2015.

[Spi16] Spille, C. *Characteristation of Reactive Systems by Means of a SuperFocus-Mixer.* master thesis, 2016.

[Stö09] Stöhr, M., Schanze, J. and Khalili, A. *Visualization of gas–liquid mass transfer and wake structure of rising bubbles using pH-sensitive PLIF. Experiments in fluids*, 47(1):135–143, 2009.

[Str17] Strassl, F., Grimm-Lebsanft, B., Rukser, D., Biebl, F., Biednov, M., Brett, C., Timmermann, R., Metz, F., Hoffmann, A., Rübhausen, M. and Herres-Pawlis, S. *Oxygen activation by copper complexes with an aromatic bis (guanidine) ligand. European Journal of Inorganic Chemistry*, 2017(27):3350–3359, 2017.

[Ter69] Teramoto, M., Nagayasu, T., Matsui, T., Hashimoto, K. and Nagata, S. *Selectivity of consecutive gas-liquid reactions. Journal of Chemical Engineering of Japan*, 2(2):186–192, 1969.

[Ter70] Teramoto, M., Fujita, S., Kataoka, M., Hashimoto, K. and Nagata, S. *Effect of Bubble Size on the Selectivity of Consecutive Gas-Liquid Reactions. Journal of Chemical Engineering of Japan*, 3(1):79–82, 1970.

[Tho81] Thompson, R.C. *Catalytic decomposition of peroxymonosulfate in aqueous perchloric acid by the dual catalysts silver (1+) and peroxydisulfate (2-) and by cobalt (2+). Inorganic Chemistry*, 20(4):1005–1010, 1981.

[Tim16] Timmermann, J. *Influence of bubble bouncing on mass transfer and chemical reaction. Chem. Eng. Tech.*, 2016.

[Tom98] Tomiyama, A., Kataoka, I., Zun, I. and Sakaguchi, T. *Drag coefficients of single bubbles under normal and micro gravity conditions. JSME International Journal Series B Fluids and Thermal Engineering*, 41(2):472–479, 1998.

[Tom02] Tomiyama, A., Celata, G., Hosokawa, S. and Yoshida, S. *Terminal velocity of single bubbles in surface tension force dominant regime. International Journal of Multiphase Flow*, 28(9):1497–1519, 2002.

[Uey79] Ueyama, K. and Miyauchi, T. *Properties of recirculating turbulent two phase flow in gas bubble columns. AIChE Journal*, 25(2):258–266, 1979.

[Van05] Vandu, C., Liu, H. and Krishna, R. *Mass transfer from Taylor bubbles rising in single capillaries. Chemical Engineering Science*, 60(22):6430–6437, 2005.

[Vas02] Vasconcelos, J.M., Orvalho, S.P. and Alves, S.S. *Gas–liquid mass transfer to single bubbles: effect of surface contamination. AIChE journal*, 48(6):1145–1154, 2002.

[Váz00] Vázquez, G., Cancela, M., Riverol, C., Alvarez, E. and Navaza, J. *Application of the Danckwerts method in a bubble column: effects of surfactants on mass transfer coefficient and interfacial area. Chemical Engineering Journal*, 78(1):13–19, 2000.

[VB04] Van Baten, J. and Krishna, R. *CFD simulations of mass transfer from Taylor bubbles rising in circular capillaries. Chemical Engineering Science*, 59(12):2535–2545, 2004.

[War10] Warnecke, H.J., Bothe, D., Zrenner, A., Berth, G. and Hüsch, K.P. *Modellbasierte Bestimmung lokal gültiger Kinetiken chemischer Reaktionen in Flüssigphase mittels Flachbettmikroreaktor. Chemie Ingenieur Technik*, 82(3):251–258, 2010.

[Was87] Wasowski, T. and Blaß, E. *Wake-Phänomene hinter festen und fluiden Partikeln. Chemie Ingenieur Technik*, 59(7):544–555, 1987.

[Whi62] White, E. and Beardmore, R. *The velocity of rise of single cylindrical air bubbles through liquids contained in vertical tubes. Chemical Engineering Science*, 17(5):351–361, 1962.

[Win66] Winnikow, S. and Chao, B. *Droplet motion in purified systems. The Physics of Fluids*, 9(1):50–61, 1966.

[Zeh85] Zehner, P. *Beschreibung der Fluiddynamik von gleichmässig fluidisierten Kugelschwärmen. Chemical Engineering and Processing: Process Intensification*, 19(1):57–65, 1985.

[Zeh88] Zehner, P. *Modellbildung für Mehrphasenströmungen in Reaktoren. Chemie Ingenieur Technik*, 60(7):531–539, 1988.

[Zub65] Zuber, N. and Findlay, J. *Average volumetric concentration in two-phase flow systems. Journal of heat transfer*, 87(4):453–468, 1965.

Experimental Analysis of Fast Reactions in Bubbly Flows

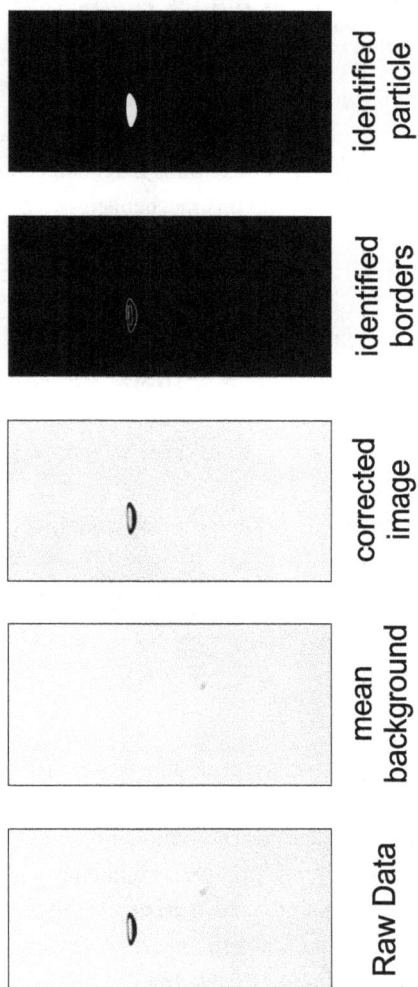

Figure 1 Flow chart of the image processing for the determination of bubble shape, velocity and trajectory

Reaction Mechanism of the Sodium Sulfite Oxidation

Reaction Mechanism according to Connick *et al.*

In 1995 Connick *et al.*[Con95] postulated a reaction mechanism for an uncatalyzed reaction (see figure 2). Nevertheless, the influence of the pH value is as well not included. Additionally, it should be noted, that the investigation of a uncatalyzed reaction is tough, since extraordinary clean substances and experimental work is necessary to confirm this mechanism.

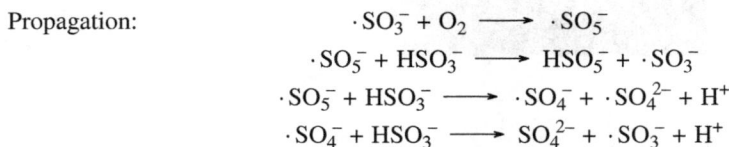

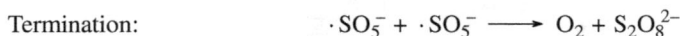

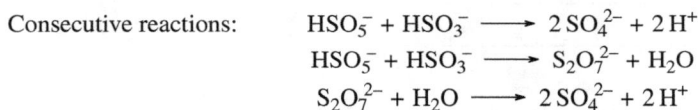

Initiation:
$$HSO_5^- + HSO_3^- \longrightarrow \cdot SO_4^- + \cdot SO_3^- + H_2O$$

Propagation:
$$\cdot SO_3^- + O_2 \longrightarrow \cdot SO_5^-$$
$$\cdot SO_5^- + HSO_3^- \longrightarrow HSO_5^- + \cdot SO_3^-$$
$$\cdot SO_5^- + HSO_3^- \longrightarrow \cdot SO_4^- + \cdot SO_4^{2-} + H^+$$
$$\cdot SO_4^- + HSO_3^- \longrightarrow SO_4^{2-} + \cdot SO_3^- + H^+$$

Termination:
$$\cdot SO_5^- + \cdot SO_5^- \longrightarrow O_2 + S_2O_8^{2-}$$

Consecutive reactions:
$$HSO_5^- + HSO_3^- \longrightarrow 2\,SO_4^{2-} + 2\,H^+$$
$$HSO_5^- + HSO_3^- \longrightarrow S_2O_7^{2-} + H_2O$$
$$S_2O_7^{2-} + H_2O \longrightarrow 2\,SO_4^{2-} + 2\,H^+$$

Figure 2 Reaction mechanism according to Connick *et al.*[Con95].

While Bäckström could not reinforce his reaction mechanism with experimental data of the sulfite oxidation, the mechanism of Connick *et al.* is proofed with findings of different workgroups. With the use of pulse radiolysis and flash photolysis the chain carrier was identified. It is proven, that $\cdot SO_3^-$ is formed in sulfite solutions and is reacting fast with oxygen to $\cdot SO_5^-$.

In difference to the postulated reaction of Bäckström

$$\cdot SO_5^- + HSO_3^- \longrightarrow HSO_5^- + \cdot SO_3^-$$

Connick *et al.* stated, that this step alone does not clarify the inhibition of the reaction in presence of alcohols. Indeed Dogliotti and Hayon [Dog67] found already, that neither the radicals $\cdot SO_5^-$ nor $\cdot SO_3^-$ react easily with alcohols, but that a fast quenching of $\cdot SO_4^-$ occurs. Therefore they postulated

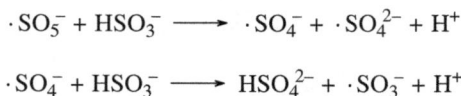

$$\cdot SO_5^- + HSO_3^- \longrightarrow \cdot SO_4^- + \cdot SO_4^{2-} + H^+$$

$$\cdot SO_4^- + HSO_3^- \longrightarrow HSO_4^{2-} + \cdot SO_3^- + H^+$$

However it was still unclear if the reaction of $\cdot SO_5^-$ with HSO_3^- proceeds according to the proposed reaction step by Bäckström or Dogliotti and Hayon. Connick *et al.* proposed, based on data of different work groups, that both reaction pathway are progressing simultaneously. On one hand it was proven that SO_5^{2-} exists in basic solutions. On the other hand, an intermediate product was found, $S_2O_7^{2-}$, which is formed by the reaction of peroxymonosulfate anion with a bisulfite ion.

There are several different reactions that could be chain terminating, where the chain carriers are reacting with each other or itself.

$$\cdot SO_3^- + \cdot SO_3^- \longrightarrow SO_3^{2-} + SO_3$$

$$\cdot SO_3^- + \cdot SO_3^- \longrightarrow S_2O_6^{2-}$$

$$\cdot SO_5^- + \cdot SO_5^- \longrightarrow O_2 + S_2O_8^{2-}$$

$$\cdot SO_3^- + \cdot SO_5^- \longrightarrow S_2O_8^{2-}$$

It is shown, that the depletion of $\cdot SO_5^-$ and $\cdot SO_3^-$ is fast, but the reaction rate of he last reaction above is not revealed yet, therefore this step is not included in the mechanism of Connick *et al.*. Based on the studies of Thompson [Tho81], who observed that the reaction kinetics for $\cdot SO_5^-$ to oxygen and $S_2O_8^{2-}$ is a factor of ten smaller than the reaction to $\frac{1}{2} O_2$ and $\cdot SO_4^-$, these reactions are added to the mechanism. Although Connick *et al.* stated, that beside the described reactions also additional reactions are possible that could be chain terminating. For example the reaction of a chain carrier with a free radical.

Furthermore, the process of initiation is still unclear, so that a photochemical or thermal initiation are possible. Nevertheless within literature it is assumed, that there is no uncatalyzed reaction and the initiation proceeds through small amounts of metal ions [Con95].

Reaction Mechanism according to Ermakov und Purmal

Ermakov and Purmal published in 2001 an article that describes the catalytic sulfite oxidation through traces of iron ions (see figure 3). It is stated, that the catalysis is also present in nearly every case and experimental data from various publications an iron concentration in the range of 10^{-8} bis 10^{-9} $mol \cdot L^{-1}$ is calculated. Due to these iron ion traces the bad reproducibility of most uncatalyzed reactions can be explained. Additionally also the high dependency of the reaction rate from the ph value can be described by traces of iron ions. For example at a pH value below 3, Fe^{3+} ions are hydrolyzed

$$Fe^{3+} + H_2O \rightleftharpoons FeOH^{2+} + H^+$$

The concentration of $FeOH^{2+}$ is therefore higher. Within the proposed mechanism, $Fe(OH)SO_3H^+$ is responsible for chain initiation which is formed by the reaction of HSO_3^- with $FeOH^{2+}$. Due to a rising concentration of $FeOH^{2+}$ a higher reaction rate results. For pH values above 4, $FeOH^{2+}$ is hydrolyzed to $Fe(OH)_2^+$

$$FeOH^{2+} + H_2O \rightleftharpoons Fe(OH)_2^+ + H^+$$

Therefore, the concentration of $FeOH^{2+}$ is decreasing and a slower reaction rate results.

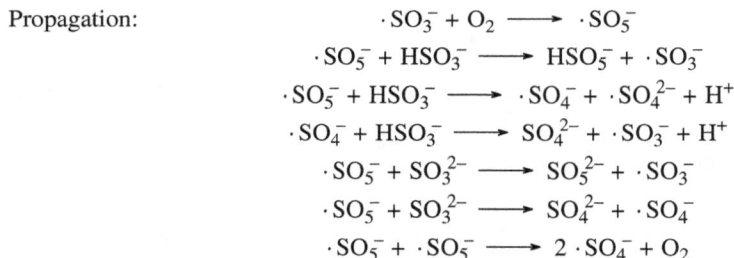

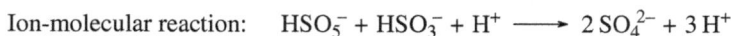

Initiation: $\qquad Fe(OH)SO_3H^+ \longrightarrow Fe^{2+} + \cdot SO_3^- + H_2O$

Propagation:
$$\cdot SO_3^- + O_2 \longrightarrow \cdot SO_5^-$$
$$\cdot SO_5^- + HSO_3^- \longrightarrow HSO_5^- + \cdot SO_3^-$$
$$\cdot SO_5^- + HSO_3^- \longrightarrow \cdot SO_4^- + \cdot SO_4^{2-} + H^+$$
$$\cdot SO_4^- + HSO_3^- \longrightarrow SO_4^{2-} + \cdot SO_3^- + H^+$$
$$\cdot SO_5^- + SO_3^{2-} \longrightarrow SO_5^{2-} + \cdot SO_3^-$$
$$\cdot SO_5^- + SO_3^{2-} \longrightarrow SO_4^{2-} + \cdot SO_4^-$$
$$\cdot SO_5^- + \cdot SO_5^- \longrightarrow 2 \cdot SO_4^- + O_2$$

Termination: $\qquad SO_5^- + \cdot SO_5^- \longrightarrow O_2 + S_2O_8^{2-}$

Ion-molecular reaction: $\quad HSO_5^- + HSO_3^- + H^+ \longrightarrow 2\,SO_4^{2-} + 3\,H^+$

Figure 3 Reaction mechanism according to Ermakov und Purmal [Erm01].

The reaction mechanism is divided in ten steps, while most steps are adopted from other work groups and not all are supported by experimental data. But the mechanism is very similar to the one reported by Connick *et al.*[Con95] except the initiation step. Additionally more propagation steps are discussed, but Connick *et al.* already stated that more propagation steps are likely. These steps are combinations of found educts and products within the solution and are plausible, but the degree of complexity for a description is rising with every step [Erm01].

Supervised Theses

This work includes experimental data and results of the following supervised student theses:

- *K. Rudnik,* Investigation on the influence of bubble collisions on mass transfer into the bulk phase, master theses, 2015

- *C. Spille,* Determination of microkinetics by investigations on local concentration fields within a SuperFocus mixer, project theses, 2015

- *R. Wank,* Experimental investigation on the influence of bubble collisions on physical mass transfer, bachelor theses, 2016

- *J. Grewe,* Investigations on the mass transfer a freely rising oxygen bubbles with and without chemical reaction, bachelor theses, 2016

- *C. Spille,* Characterization of reactive systems by means of a SuperFocus Mixer, master theses, 2016

- *M. Döpken,* Influence of the flow regime on chemical reactions in bubble columns, master theses, 2017

- *C. Compart,* Design of an experimental setup for the analysis of mass transfer on reactive Taylor bubbles, master theses, 2017

Lebenslauf

Name	Timmermann
Vorname	Jens
Geburtsdatum	31.03.1987
Geburtsort	Höxter
Geburtsland	Deutschland

08/1993 - 07/1997	Katholische Grundschule der Stadt Brakel
08/1997 - 07/2003	Annette-von-Droste-Hülshoff-Realschule Brakel
08/2003 - 06/2006	Ausbildung zum Chemikanten bei Symrise GmbH & Co. KG in Holzminden
06/2006 - 08/2006	Chemikant bei Symrise GmbH & Co. KG in Holzminden
09/2006 - 06/2007	Fachhochschulreife am Adolf-Kolping-Berufskolleg des Kreises Höxter in Brakel
10/2007 - 10/2010	Bachelorstudium Chemie an der Universität Paderborn, Abschluss: Bachelor of Science
10/2010 - 11/2012	Masterstudium Chemie an der Universität Paderborn, Abschluss: Master of Science
12/2012 - 12/2017	wissenschaftlicher Mitarbeiter am Institut für Mehrphasenströmungen der Technischen Universität Hamburg
01/2018 - 09/2018	Anfertigung der Dissertation am Institut für Mehrphasenströmungen der Technischen Universität Hamburg
seit 10/2018	Process Optimization Specialist bei der Sasol Germany GmbH

Publications

Paper:

Hermann, P., Timmermann, J., Hoffmann, M., Schlüter, M., Hofmann, C., Löb, P., & Ziegenbalg, D. (2018). Optimization of a split and recombine micromixer by improved exploitation of secondary flows. Chemical Engineering Journal, 334, 1996-2003. DOI: 10.1016/j.cej.2017.11.131

Kastens, S., Timmermann, J., Strassl, F., Rampmaier, R. F., Hoffmann, A., Herres-Pawlis, S., & Schlüter, M. (2017). Test System for the Investigation of Reactive Taylor Bubbles. Chemical Engineering & Technology, 40(8), 1494-1501. DOI: 10.1002/ceat.201700047

Sellin, D., Hiessl, R., Bothe, M., Timmermann, J., Becker, M., Schlüter, M., & Liese, A. (2017). Simultaneous local determination of mass transfer and residence time distributions in organic multiphase systems. Chemical Engineering Journal, 321, 635-641. DOI: 10.1016/j.cej.2017.03.150.

Mierka, O., Munir, M., Spille, C., Timmermann, J., Schlüter, M., & Turek, S. (2017). Reactive Liquid-Flow Simulation of Micromixers Based on Grid Deformation Techniques. Chemical Engineering & Technology, 40(8), 1408-1417. DOI: 10.1002/ceat.201600686

Timmermann, J., Hoffmann, M., & Schlüter, M. (2016). Influence of Bubble Bouncing on Mass Transfer and Chemical Reaction. Chemical Engineering & Technology, 39(10), 1955-1962. DOI: 10.1002/ceat.201600299

Others:

Strassl, F., Timmermann, J., Schlueter, M., Herres-Pawlis, S. (2016). Kinetik der Sauerstoffaktivierung. GIT Labor-Fachzeitschrift. 39-41

Oral Presentations:

Jahrestreffen der Fachgruppen Computational Fluid Dynamics und Mehrphasenströmungen, 2015, Lueneburg, Germany: "Einfluss von Blasenkollisionen in Blasenströmungen auf Stofftransport und chemische Reaktion", Jens Timmermann, Marko Hoffmann, Michael Schlueter

Jahrestreffen der ProcessNet-Fachgruppen Agglomerations- und Schüttguttechnik, Mehrphasenströmungen und Computational Fluid Dynamics, 2016, Bingen, Germany: "Experimentelle Analyse der Grenzschichtdynamik in Blasenströmungen", Jens Timmermann, Marko Hoffmann, Michael Schlueter

9th International Conference on Multiphase Flows, 2016, Florence, Italy: "Influence of bubble bouncing on mass transfer and chemical reaction", Jens Timmermann, Marko Hoffmann, Michael Schlueter

Symposium on Non-Invasive Measuring Tools and Numerical Methods for the Investigation of Non-Reactive and Reactive Gas-Liquid Flows, FERMaT-SPP1740 Symposium Toulouse 2016, France: "Influence of bubble bouncing on mass transfer and chemical reaction", Jens Timmermann, Marko Hoffmann, Michael Schlueter

Jahrestreffen Dresden - Jahrestreffen der ProcessNet-Fachgruppen Mehrphasenströmungen, Partikelmesstechnik, Zerkleinern und Klassieren, Computational Fluid Dynamics, Mischvorgänge und dem TAK Aerosoltechnologie, 2017, Dresden, Germany: "Einfluss der Grenzschichtdynamik auf den Stofftransport und chemische Reaktionen", Jens Timmermann, Marko Hoffmann, Michael Schlueter

Poster Presentations:

ESCRE 2015 - European Symposium on Chemical Reaction Engineering, 2015, Fuerstenfeldbruck, Germany: "Experimental analysis of boundary layer dynamics inbubbly flows", Jens Timmermann, Marko Hoffmann, Michael Schlueter

MMPE2017 - Third International Symposium on Multiscale Multiphase Process Engineering, 2017, Toyama, Japan: "Influence of boundary layer deformations on mass transfer and chemical reaction", Jens Timmermann, Marko Hoffmann, Michael Schlueter

www.ingramcontent.com/pod-product-compliance
Lightning Source LLC
Chambersburg PA
CBHW081109220326
41598CB00038B/7289